Catecholamine Research in the 21st Century

Abstracts and Graphical Abstracts, 10th International Catecholamine Symposium, 2012

Catecholamine Research in the 21st Century

Abstracts and Graphical Abstracts,
10th International Catecholamine
Symposium, 2012

Lee E. Eiden

AMSTERDAM • BOSTON • HEIDELBERG • LONDON
NEW YORK • OXFORD • PARIS • SAN DIEGO
SAN FRANCISCO • SINGAPORE • SYDNEY • TOKYO
Academic Press is an imprint of Elsevier

Academic Press is an imprint of Elsevier
32 Jamestown Road, London NW1 7BY, UK
225 Wyman Street, Waltham, MA 02451, USA
525 B Street, Suite 1800, San Diego, CA 92101-4495, USA

Notice

No responsibility is assumed by the publisher for any injury and/or damage to persons or property
as a matter of products liability, negligence or otherwise, or from any use or operation of any
methods, products, instructions or ideas contained in the material herein.

Because of rapid advances in the medical sciences, in particular, independent verification of
diagnoses and drug dosages should be made

British Library Cataloguing-in-Publication Data
A catalogue record for this book is available from the British Library

Library of Congress Cataloging-in-Publication Data
A catalog record for this book is available from the Library of Congress

ISBN: 978-0-12-800044-1

For information on all Academic Press publications
visit our website at elsevierdirect.com

Typeset by MPS Limited, Chennai, India
www.adi-mps.com

Printed and bound by CPI Group (UK) Ltd, Croydon, CR0 4YY
Transferred to digital print 2012

Working together
to grow libraries in
developing countries

www.elsevier.com • www.bookaid.org

Contents

THEME A   CATECHOLAMINE BIOSYNTHESIS AND STORAGE

THEME B   CATECHOLAMINE RELEASE AND RE-UPTAKE

THEME C METABOLISM

THEME D   CATECHOLAMINE RECEPTORS AND CATECHOLAMINERGIC SIGNALING

THEME E   NEUROLOGY

THEME F   PSYCHIATRY AND PSYCHOLOGY

THEME G   DRUG ABUSE AND ADDICTION

THEME H   CATECHOLAMINES IN THE PERIPHERY

THEME I   CATECHOLAMINES IN INTEGRATIVE FUNCTION

THEME J   CATECHOLAMINE INTERACTIONS WITH OTHER TRANSMITTERS

Preface

The Tenth International Catecholamine Symposium (XICS) was held at The Asilomar Conference Grounds in Pacific Grove CA, September 9-13, 2012. It was the first international symposium focused on catecholamines in the 21st century, providing the title for the proceedings of the symposium found in this volume. David Goldstein, founding Chief of the Clinical Neurocardiology Section, NINDS and President of the 8th International Catecholamine Symposium, also held at Asilomar in 1996, provided the guiding inspiration for the XICS. Daniel O'Connor, Professor of Medicine and Pharmacology at the Institute for Genomic Medicine, University of California San Diego, and President of the Catecholamine Society was also President of the 10th International Catecholamine Symposium. David Sibley, Chief of the Molecular Neuropharmacology Section, NINDS, Bethesda, MD, Esther Sabban, Professor of Biochemisty and Molecular Biology, New York Medical College, and the editor of this volume constituted the XICS Executive Organizing Committee. Many others, most especially the participants who came from Austria, Canada, The Czech Republic, Denmark, France, Germany, Israel, Italy, Japan, Mexico, Norway, Russian, Taiwan, the United States and elsewhere, contributed importantly to the success of the XICS. The editor wishes to thank Dave Goldstein and Dan O'Connor especially for their devotion to the eventual completion of the Symposium and these proceedings.

Some of the broader topics most relevant to the current status of catecholamine research as a translational field have been covered in greater depth, by thought leaders in catecholamine research who were in attendance at the XICS, in a companion volume of Advances in Pharmacology. These include mechanisms of catecholamine biochemistry, cell biology, systems biology, clinical diagnosis, drug discovery and target discovery, and gene therapeutic treatment for catecholamine-related human disease. Contained herein are the comprehensive conference proceedings in the form of extended graphical abstracts, and some condensed abstracts, of almost all of the presentations at the meeting organized into the ten themes under which the symposium was convened. The themes are introduced as chapters each with a short introduction that attempts to identify the highlights and 'growth areas' within each. It is hoped that the reader will find these proceedings useful as a handbook for the current state of play of catecholamine research, until such time as the Proceedings of the 11th International Catecholamine Symposium succeeds them.

Lee E. Eiden

Section on Molecular Neuroscience
Laboratory of Cellular and Molecular Regulation
National Institutes of Health
Bethesda, MD, USA

Catecholamine Biosynthesis and Storage

Lee Eiden and David Goldstein

The major features of catecholamine biosynthesis [Tyrosine (TH) –> L-Dopa (AADC) –> DA (DBH) –> NE (PNMT) –> Epi] and storage [cytoplasmic DA, NE, Epi –> VMAT1,2 –> vesicular DA, NE, Epi] were worked out well before the end of the previous century. This chapter contains illustrative examples of progress since, and this is manifest in a more complete molecular understanding of the cell biology that allows TH to control CA biosynthesis; concrete clinical steps for detection and gene therapy for catecholamine deficiency diseases; and complete identification of the combinations of catecholamine biosynthetic enzymes and vesicular storage capacity that can be found throughout the brain and periphery, and distinctly in rodent and primate neuronal systems, from which a 'post-classical' view of catecholamine chemical neuroanatomy has emerged.

While TH retains pride of place as the rate limiting enzyme for catecholamine biosynthesis, new insights into the conversion of tyrosine to L-Dopa have been gained by fuller exploration of the role of GTP cyclohydrolase (GTPCH) in supplying the necessary co-factor, tetrahydrobiopterin, for this enzymatic conversion. GTPCH and TH converge biochemically on tyrosine to allow its conversion to

Catecholamine Research in the 21st Century.

L-Dopa, and thus share in this rate-limiting step for catecholamine production. This provides fundamental molecular insights and clinical opportunities. One of the latter is that GTPCH deficiency occurs in humans, causes disease, and can potentially be corrected.

Enzymes in a metabolic pathway are non-rate-limiting when their increase does not increase turnover in the pathway: any enzyme *becomes* rate-limiting when its decrease or mis-trafficking cause its abundance to be less than the previous 'rate-limiting enzyme' in the pathway. Deficiency of AADC causes motor impairment that can be corrected by expression of AADC via viral vector-mediated gene delivery directly to the brain.

Insights into how different types of catecholaminergic neurons develop in the brain have indexed progress in developmental biology as a field. A surprising new insight is that the brain *during* development may itself be an endocrine organ secreting catecholamines to the rest of the body. The presumption of developmental neuroscientists that 'TH-positive' neurons are dopaminergic or noradrenergic has been shattered in the last decade: there are DOPAergic and trace aminergic neurons, as well as neuronal dyads that synthesize dopamine only via intercellular collaboration. The notion that the chemical neuroanatomy of the human nervous system can be fully comprehended by study of the rodent nervous system has likewise been rendered untenable, and translational neuroscience is far better for it. Finally, exciting new vistas have emerged from the use of genetically-based lesioning and complementation experiments that show that subdivisions of the major catecholamine nuclei of the brain, including the locus coeruleus and substantia nigra, have surprising heterogeneity of projections with specific and distinct functions.

Is further progress mainly a matter of effectively executing therapeutic strategies, or do further questions remain? They do. It is still unclear for example how supply of AADC to cells of the striatum corrects a deficiency that inheres primarily in neurons that *project* to the striatum. The view that imaging of VMAT2 with TBZ reflects the molar concentration of the protein in the brain has given way to an emerging understanding that endogenous catecholamine levels play a role in how much TBZ binds to VMAT2 in the brains of normals, addicts to methamphetamine or cocaine, and patients with progressive degenerative disorders. The near-complete success of virus-mediated correction of brain defects related to catecholamine biosynthesis, and pharmacological correction of genetic deficiencies in catecholamine storage (see for example Rilstone et al., N. Engl. J. Med. 368, 543, 2013) show that robust clinical gains have resulted from translation of pre-2000 information, tools, and advances in allied fields. This in turn predicts that gains in new knowledge reported in this first, as well as the following chapters of "Catecholamine Research in the 21st Century" will likely parlay into further therapeutic gains in the coming decade.

Genetic Manipulation of Catecholamine Signaling in the Mouse

Richard Palmiter

University of Washington School of Medicine Seattle, Washington, USA

Gene targeting in mouse embryonic stem cells has been used to generate mice that are unable to synthesize each of the major catecholamines. The developmental and behavioral consequences of making mice that are unable to synthesize epinephrine, norepinephrine and dopamine will be discussed. I will illustrate how restoring dopamine signaling to specific brain regions of an otherwise dopamine-deficient mouse can restore viability and the ability to engage in specific behaviors. I will also show how we manipulate the activity of dopamine-producing neurons to affect the behavior of mice.

AADC Deficiency: Occurring in Humans; Modeled in Rodents; Treated in Patients

Wuh-Liang Hwu[1], Ni-Chung Lee[1], Yih-Dar Shieh[1], Kai-Yuan Tzen[2], Pin-Wen Chen[1], Shin-ichi Muramatsu[3], Hiroshi Ichinose[4] and Yin-Hsiu Chien[1]

[1]Department of Medical Genetics, Pediatrics; [2]Nuclear Medicine, National Taiwan University Hospital; [3]Division of Neurology, Department of Medicine, Jichi Medical University, Japan; [4]Department of Life Science, Graduate School of Bioscience and Biotechnology, Tokyo Institute of Technology, Japan

Aromatic L-amino Acid Decarboxylase (AADC) deficiency (MIM #608643) is an autosomal recessive inborn error of neurotransmitter metabolism which causes severe motor dysfunction, oculogyric crisis, autonomic dysfunction, and emotional liability in patients since infancy. AADC deficiency is more common in Taiwan than in other countries because of a foundermutation (IVS6 + 4A>T). We have established a knock-in (KI) mouse model (DdcIVS6/IVS6) for AADC deficiency. Some of the homozygous KI mice were born alive, but they exhibited severe failure to thrive, dyskinesia, and clasping. However, if they could survive their first few weeks of lives, they then caught up in growth and improved in motor function. Gene therapy at neonatal stage could eliminate the manifestations of the disease. An attempt to treat patients with AADC deficiency was initiated before the establishment of the mouse model. AAV2-AADC vectors were infused into the bilaterally putamen of four patients 4 to 6 years of age. All of the patients showed improvements in motor performance: one patient was able to stand 16 months after gene transfer, and the other three patients gained head controls or achieved supported sitting. Choreic dyskinesia was observed in all patients, but this resolved after several months. 6-[18F]fluorodopa positron emission tomography (PET) and cerebrospinal fluid analysis both showed evidences of the treatment effects.

Tyrosine Hydrolylase and Dopamine Beta-Hydroxylase: Role of Common Genetic Variation in Adrenergic Responses to Stress and in Hypertension

Daniel O'Connor
UCSD, USA

RATIONALE: Catecholamine biosynthesis is catalyzed by a pathway of reactions in series. We asked whether the substantial inter-individual variation in catecholaminergic responses was in part heritable, and referable to genetic variation at loci encoding such enzymes. We also explored implications of pathway genetic variation for catecholamine secretion, environmental stress responses, and disease.

METHODS: We phenotyped catecholaminergic responses in twin pairs (MZ and DZ) to estimate trait heritability (h2), and re-sequenced the tyrosine hydroxylase (TH) and dopamine beta-hydroxylase (DBH) loci in n = 80 individuals (i.e., 2n = 160 chromosomes) each for systematic polymorphism discovery. We explored whether such variants predicted catecholamine secretion, BP responses to environmental stress, or hypertension in the population. Finally, we tested chromaffin cell-transfected luciferase reporter plasmids for consequences of promoter variation.

RESULTS: Both catecholamine secretion and environmental stress (BP response to cold) traits were substantially heritable as judged by twin pair variance components. Most of the variation at TH and DBH occurred as bi-allelic SNPs (single nucleotide polymorphisms). Variations within the open reading frame (ORF), i.e., non-synonymous (amino acid replacement) changes, were unusual. By contrast, functional, trait-associated variation was localized to the promoter regions of both TH and DBH.

At TH, 4 SNPs in the proximal promoter gave rise to haplotypes predicting catecholamine secretion and cold-stress BP change. One of these SNPs predicted BP in the population. Two of these promoter SNPs (C-824T, rs10770141, frequency ~39%; A-581G, rs10770140, frequency ~38%) differentially influenced gene expression in TH promoter haplotype/luciferase reporter plasmids transfected into chromaffin cells. At DBH, 6 SNPs in the proximal promoter gave rise to haplotypes predicting DBH secretion in vivo. One of these promoter SNPs (C-970T, rs1611115; C-2073T, rs1989787) predicted plasma DBH activity, catecholamine secretion, stress BP changes, and basal BP in populations of European and African ancestry. The same SNP differentially influenced gene expression in DBH promoter haplotype/luciferase reporter plasmids transfected into chromaffin cells.

CONCLUSIONS: Naturally occurring, common genetic variations at TH andDBH, especially in the proximal promoter regions, have functional consequences, and predispose to cardio-renal disease risk.

ACKNOWLEDGMENTS: Critical participants in these studies, without whom the work would not have been possible, include Lian Zhang, Yuqing Chen ,Gen Wen, Kuixing Zhang, Fangwen Rao, Laurent Taupenot, Sushil K. Mahata, Manjula Mahata, Sajalendu Ghosh, Brinda K. Rana, Jennifer Wessel, Maple M. Fung, Michael G. Ziegler, Nicholas J. Schork, and Bruce A. Hamilton.

Pharmacokinetic and Pharmacodynamic Properties of Etamicastat, a New DBH Inhibitor: Comparison to Nepicastat

Ana I. Loureiro[1], Maria João Bonifácio[1], Carlos Fernandes-Lopes[1], Bruno Igreja[1], Nuno Pires[1], Lyndon Wright[1] and Patrício Soares-da-Silva[1,2]

[1]Dept. Research & Development, BIAL — Portela & Cª, S.A., 4745-457 S. Mamede do Coronado, Portugal; [2]Dept. Pharmacology & Therapeutics, Faculty of Medicine, University Porto, Porto, Portugal.

BACKGROUND: Dopamine-β-hydroxylase (DBH) catalyses the conversion of dopamine (DA) to norepinephrine (NE) in the sympathetic nervous system and is considered a potential therapeutic target in hypertension and chronic heart failure. Nepicastat and etamicastat are DBH inhibitors that decrease NE levels in sympathetically innervated tissues and slow the drive of the sympathetic nervous system.

OBJECTIVES: This study compared mouse catecholamine level modulation, DBH inhibition and exposure following a single oral dose of nepicastat and etamicastat.

METHODS AND RESULTS: Nepicastat and etamicastat were orally administered to NMRi mice and DBH activity, catecholamine levels and compounds biodisposition were evaluated. Following administration, nepicastat and etamicastat produced a NE reduction in heart of about 60 % and 40 % and a dopamine increase of 1000% and 800%, respectively; nevertheless the NE/DA ratios in the heart were similar after nepicastat and etamicastat. Only nepicastat altered catecholamine levels in the frontal and parietal cortexes with decreases of NE levels by 60% of control values. The overall exposure to nepicastat (33.0 ± 7.2 h.μg/ml) was 8-fold that of etamicastat (4.3 ± 0.6 h.μg/ml). In brain only nepicastat (7.6 ± 0.9 h.μg/ml) was detected, which may explain the decrease in NE levels in frontal and parietal cortexes after nepicastat. For etamicastat the maximum DBH inhibition was obtained at 1 h with complete recovery after 15 h. When nepicastat was administered DBH maximum inhibition was observed from 1 to 9 h and enzymatic recovery was observed after 24 h.

CONCLUSIONS: Nepicastat and etamicastat are potent and orally active inhibitors of DBH leading to gradual modulation of the sympathetic nervous system by inhibiting the biosynthesis of NE. Etamicastat, as a peripheral DBH inhibitor, is a good candidate for the treatment of cardiovascular disorders associated with over-activation of the sympathetic nervous system, without adverse effects related to changes in brain catecholamine levels.

Table 1 Mean NE/DA Tissue Ratios Following *p.o.* Administration of 30 mg/kg Nepicastat and Etamicastat

Time (h)	Etamicastat NA/DA Mean ± SEM	Nepicastat NA/DA Mean ± SEM	n
Heart			
Pre-dose	1.02 ± 0.07	1.06 ± 0.16	5
1	0.22 ± 0.01*	0.21 ± 0.01*	5
3	0.13 ± 0.01*	0.12 ± 0.02*	5
9	0.07 ± 0.01*	0.08 ± 0.02*	5
24	0.28 ± 0.11*	0.08 ± 0.01*	5
Parietal Cortex			
Pre-dose	1.18 ± 0.19	1.47 ± 0.42	5
1	0.93 ± 0.13	0.42 ± 0.09*	5
3	1.01 ± 0.28	0.33 ± 0.04*	5
9	0.94 ± 0.13	0.19 ± 0.03*	5
24	1.07 ± 0.27	0.61 ± 0.10*	5

*Significantly different from corresponding pre-dose values ($P < 0.05$).

Estradiol-Mediated Regulation of Gene Expression of Catecholamine Biosynthetic Enzymes: The Role of Membrane-Initiated Signaling with ERα

Lidia Serova, Shreekrishna Maharjan, Regina Nostramo and Esther Sabban

New York Medical College USA

BACKGROUND: Estrogens have profound effects on brain catecholaminergic (CA) systems. However within the CA systems, there are contradictory findings regarding its effects. The discrepancies might be explained, at least partially, by estrogen receptor (ER) specific regulation of genes related to CA biosynthesis.

OBJECTIVES: Here we examined the role of ER subtypes in estrogen-mediated regulation of gene expression for TH, DBH and GTP cyclohydrolase I (GTPCH).

METHODS AND RESULTS: In vivo experiments were performed on ovariectomized (OVX) Sprague-Dawley female ratsinjected with estradiol benzoate (EB), ERα (PPT) or ERβ (DPN) agonists for different durations and then subjected to immobilization stress (IMO). During IMO, blood pressure was monitored. TH, DBH and GTPCH mRNA levels were measured in several brain areas by real-time RT-PCR. In vitro experiments were performed on PC12 cells transfected with ERα or ERβ expression vectors and either intact or mutated TH/Luc, DBH/Luc or GTPCH/Luc promoter reporter constructs. Pharmacological treatments were performed using 17 β-estradiol (E2), membrane-impermeable estradiol conjugate (E2BSA), PPT, DPN, protein kinase A (PKA) and MEK inhibitors. Western blot analysis was used to assess levels of phospho-CREB. Short-term injections of EB to OVX rats led to parallel elevation of TH, DBH and GTPCH mRNA levels in the locus coeruleus (LC) and nucleus of the solitary tract (NTS). However in the substantia nigra (SN), only GTPCH, but not TH, mRNA levels were increased, possibly due to differential distribution of ER subtypes and specificity in transcriptional regulation of these genes. Long-term injections of EB modulated basal and IMO-induced expression of these genes. For example, treatment of OVX rats with EB significantly reduced stress-elicited rise of TH and DBH mRNA levels. In parallel with reduced activation of CA neurons, these rats had also lower blood pressure response to IMO. Results with long—term injections of PPT and DPN to OVX rats showed a significant influence on the response to acute IMO, while PPT was more effective with repeated IMO implicating involvement of ERα. To study molecular mechanism of estrogen-mediated regulation of CA biosynthesis we performed experiments on PC12 cells. Depending on the ER subtype, an opposite response of the TH promoter to E2 was observed, with an increase with ERαand decrease with ERβ. However, both DBH and

GTPCH promoters were enhanced by E2 over a wide range of concentrations with either ER subtype. Membrane-initiated estradiol signaling in regulation of TH and DBH promoter activity was demonstrated using membrane-impermeable estradiol conjugate (E2BSA). E2BSA elicited rapid phosphorylation of CREB and increased CRE-driven promoter activity. Over-expression of dominant negative forms of CREB, with mutations in DNA binding or phosphorylation site, prevented TH promoter response to E2BSA. Pre-treatment with protein kinase A (PKA) and MEK inhibitors reduced E2 dependent phosphorylation of CREB and ERK, and also decreased induction of TH promoter activity by E2 or E2BSA.

CONCLUSIONS: The results in vivo demonstrate the importance of estrogens in regulation of basal and stress induced CA biosynthetic enzyme gene expression. Activation of ERα is especially important in modulation of the response to stress. The findings in vitro show that E2 induction of TH and DBH gene transcription with ERα involves membrane-initiated estradiol signaling. The role of membrane initiated ER signaling in regulation of CA gene expression by estrogens in vivo remains to be determined.

Structural Basis for Regulation of Tyrosine Hydroxlyase by the Catecholamines

Gabrielle Briggs, Jesse Bulley, Peter R. Dunkley and Phillip W. Dickson

University of Newcastle Australia

BACKGROUND: Tyrosine Hydroxylase (TH) performs the first and rate limiting step in the synthesis of catechoamines (CAs), which feedback to dramatically inhibit DOPA synthesis by irreversibly binding to the active site. Key residues in the N-terminal regulatory domain are required for this inhibition and are thought to interact with the CA-bound active site. During neuronal or chromaffin cell activation, phosphorylation of the N-terminal residue Ser40 relieves this inhibition. In addition, another CA binding site has lower affinity for the enzyme, is dissociable and is not abolished by phosphorylation. Both this site and the non-dissociable site inhibit TH by competing with the essential cofactor, tetrahydrobiopterin (BH4). In situ, the low affinity site regulates TH through binding and dissociating according to cytosolic CA levels, and operates in conjunction with the "on/off" regulation provided by high affinity CA binding and phosphorylation. While the functionality of these regulatory sites is understood, the structural basis underlying their functions is mostly unknown. This is largely due to the TH crystal structure being a CA-free form and lacking the 156 N-terminal residues containing those required for high affinity CA binding.

OBJECTIVES: We aimed to identify residues involved in the CA binding sites via substitution of amino acids within and around the active site. In doing so, we would gain insight into how TH is regulated by CAs, in the absence of a complete crystal structure. A key area identified for the high affinity site was then used for computational screening of compounds and in vitro testing of hits for their effect on high affinity CA binding.

METHODS AND RESULTS: 7 active site substitutions and 8 substitutions of residues around the outer regions of the active site were generated. Dopamine (DA) dependent inhibition of recombinant TH activity was assayed using the tritiated water release assay. The IC50s for DA inhibition through the low affinity site in active site mutants Y371F and E332D were 70-fold and 10-fold higher than wild-type respectively. Effects of CA bound to the high affinity site were assessed by measuring the competition of CA with the cofactor, BH4. High affinity bound DA produced a 10-fold increase in the Km for BH4 in wild-type TH. In Y371F and E332D, this inhibitory effect was absent. Substitutions made outside of the active site showed changes to the high affinity site, with A297L and D361N showing no increase in the Km for the cofactor upon high affinity CA binding. This region of TH was targeted for computational screening of a drug database using DOCK 6. 7 of the hits were tested in vitro. While DA inhibition was retained with these compounds, one was shown to increase TH activity up to 5-fold in the absence of DA.

CONCLUSIONS: The common roles of E332 and Y371 in low and high affinity CA binding indicate that these sites overlap in the active site and use E332 and Y371to position CA in the BH4 binding site. Residues outside the active site seem to be involved in high affinity binding only and may constitute the interacting surface for the N-terminus in producing an irreversible binding of CA. Since low and high affinity CA binding occur simultaneously, they cannot occupy the same space, as these results suggest. However, TH is a dimer of dimers and CA binding stoichiometry has previously been measured as 1mol/mol TH dimer for each of the low and high affinity sites, suggesting that E332 and Y371 bind CA in both monomers of the dimer, while A297 and D361 have an additional role in producing the high affinity site in one monomer only. This model could be confirmed by testing for high and low affinity sites in TH monomers. The compound's activation of TH activity suggests that this region may have roles outside that of high affinity CA inhibition.

Unique Regulation of TH Gene Expression in Midbrain Dopamine Neurons

William Tank

University of Rochester Medical Center, USA

BACKGROUND: In most catecholaminergic model systems stimulus-dependent induction oftyrosine hydroxylase (TH) is regulated primarily by activation of gene transcription followed by induction of

TH mRNA. The cAMP signaling pathway usually plays a major role in thistranscriptional response. However, in midbrain (MB) dopamine neurons cAMP does not induce THmRNA. Instead, cAMP induces TH protein via a post-transcriptional translational mechanismmediated by a pyrimidine-rich domain in TH mRNA 3'UTR and in part by poly-C binding proteins.

OBJECTIVES: The goals of the present study were to determine whether cAMP activates THgene transcription in MB dopamine neurons and to investigate the mechanisms for the lack oftranscriptional response to cAMP.

METHODS: Quantitative RT-PCR was used to measure changes in TH mRNA and TH primary transcripts in midbrain slice explant cultures from 10-14 day old rat pups, brain regions from adultrats or from catecholaminergic cell lines. TH mRNA half-life was measured by pulse-chasemethodology using thiouridine or inhibition of transcription with DRB. Chromatin immunoprecipitation (ChIP) assays were performed using standard procedures.

RESULTS: Midbrain slice explant cultures and MN9D cells were treated with 0.5 mM 8-CPTcAMP for different times up to 48 hr. Neither TH mRNA, nor TH RNA primary transcript-sencoding sequences within the second intron were induced at any time point. In contrast, both THmRNA and TH primary transcripts were induced in rat pheochromocytoma and mouse locuscoeruleus-derived cath.a cells. These results suggest that cAMP does not activate TH transcription in MB dopamine neurons and that this lack of response is unique. The cAMP signaling pathway isintact in MN9D cells, since cAMP treatment is associated with phosphorylation of CREB and induces mRNAs encoding VIP and NPY. ChIP assays demonstrate that CREB is bound to the THpromoter in both MN9D cells and rat pheochromocytoma cells. Interestingly, the transcriptionalresponse of the TH gene to cAMP is observed in MN9D cells that were pretreated for 1.5 hr with histone deacetylase (HDAC) inhibitors. Finally, we demonstrate that the ratio of TH primary transcripts normalized to TH mRNA is dramatically less in adult MB compared to locus coeruleusor adrenal medulla. This low ratio of TH primary transcripts/TH mRNA is also seen in MN9Dcells compared to rat pheochromocytoma or cath.a cells. Half-life analyses indicate that TH mRNAdegradation rates are much slower in MB models than in other catecholaminergic cell types.

CONCLUSIONS: Our results indicate that treatment of MB models with cAMP analog is notassociated with activation of TH gene transcription. This lack of response is unique to MB, sincecAMP activates the TH gene in other catecholaminergic cell types, and is not due to loss of thecAMP signaling pathway, since CREB phosphorylation is intact and other cAMP-responsive genesare induced in MN9D cells. Our results suggest that the lack of cAMP transcriptional response ofthe TH gene is HDAC-dependent, but more experiments are needed to understand the precise rolethat HDAC plays in regulating the MB TH gene. Finally, our results indicate that TH genetranscription rate is very slow in MB relative to other catecholaminergic cell types, but that THmRNA levels remain high in MB, because TH mRNA degradation rate is also extremely slow in these dopamine neurons. Taken together, our results indicate that TH gene expression is controlled by a unique set of mechanisms in MB. This uniqueness may be important for understanding thecontrol of dopamine levels in these neurons in relation to drug addiction, schizophrenia and Parkinson's disease.

Intracellular Stability of Tyrosine Hydroxylase: Phosphorylation and Proteasomal Digestion of the Enzyme

Akira Nakashima[1], Yoko S. Kaneko[1], Keiji Mori[1], Hiroshi Nagasaki[1], Toshiharu Nagatsu[2] and Akira Ota[1]

[1]*Department of Physiology Fujita Health University School of Medicine, Toyoake, Aichi, Japan;* [2]*Pharmacology, Fujita Health University School of Medicine, Toyoake, Aichi, Japan*

Tyrosine hydroxylase (TH) is the rate-limiting enzyme in catecholamine biosynthesis, and its N-terminus plays a critical role in controlling the catalytic activity of the enzyme. The catalytic domain is located in the C-terminal two-thirds of the molecule and binds the substrates and the cofactor. Three aromatic amino acid hydroxylases, i.e., TH, tryptophan hydroxylase, and phenylalanine hydroxylase, contain this highly conserved catalytic domain. These enzymes were reported to be substrates for the ubiquitin-conjugating enzyme system, targeting them to be degraded by proteasomes (Kojima et al., 2000; Døskeland et al., 2001, 2002). The N-terminal part of TH is supposedly located on the surface of the molecule, because TH consists of homotetramers whose subunits are gathered together by the interaction of their C-termini (Goodwill et al., 1997). Collectively, these reports suggest that the ubiquitin-proteasome pathway is a prerequisite for the degradation of TH and that the N-terminal part of TH plays a critical role in the degradation.

Recently, dysfunction in the ubiquitin-proteasome system was found to be one of major causes of the degeneration of dopaminergic neurons (McNaught et al., 2006). There is a hypothesis that dopamine (DA) quinone produced by the oxidation of an excess amount of DA promotes the loss of dopaminergic neurons following the dysfunction of their mitochondria. Therefore, it is important to examine the degradation mechanism regulating the protein level of TH, which affects DA synthesis in the cells. According to this line of evidence, we investigated the role of the N-terminal part of TH in the degradation of this enzyme.

AtT-20 mouse neuroendocrine cells, PC12D rat pheochromocytoma cells, and SH-SY5Y human neuroblastoma cells were used for the estimation of the intracellular stability and the catalytic activity of TH. These cells were transfected with a pcDNA3.1/HisA vector containing hTH1 (human TH type 1) cDNA for the expression of hTH1 proteins. For the knock-down and overexpression of 14-3-3 protein in the cells, annealed double-stranded siRNA and pFN21A expression vector containing the cDNA of 14-3-3, respectively, were used. Lactacystin and 3-methyladenine were used to inhibit ubiquitin-proteasome and autophagy-lysosome activities. TH protein, TH protein phosphorylated at its Ser residues, and 14-3-3 protein in the cells were analyzed by Western blotting and immunohistochemistry. Catecholamine amounts were measured by using HPLC coupled with an electrochemical detector.

Mutants of hTH1 lacking its N-terminal region containing the 3 phosphorylation sites (Ser19, Ser31 and Ser40) possessed high stability of the enzyme in the cells. The inhibition of proteasome activity increased the quantity of TH molecules phosphorylated at their Ser19 and Ser40, although the inhibition of autophagy activity did not affect it. These results suggest that the phosphorylation of the N-terminal portion of TH positively regulates the degradation of this enzyme through the ubiquitin-proteasome pathway. Moreover, the down-regulation of 14-3-3 proteins in the cells exogenously expressing hTH1 enhanced the stability of the enzyme. 14-3-3 protein is well known to bind to TH molecules phosphorylated at their Ser19. Therefore, 14-3-3 protein is considered to be a critical factor in regulating TH stability by acting on the N-terminus of the enzyme.

The ubiquitin-proteasome pathway is important for the degradation of TH, and the phosphorylation at the N-terminal portion of TH was shown to be a trigger for activation of this pathway.

The Peripheral Interaction of Tyrosine Hydroxylase and 14-3-3γ with Negatively Charged Phospholipidic Membranes

Anne Baumann[1], Øyvind Halskau[2] and Aurora Martinez[1]

[1]Department of Biomedicine, University of Bergen, Norway; [2]Department of Molecular Biology, University of Bergen, Norway

BACKGROUND: Tyrosine hydroxylase (TH) catalyzes the hydroxylation of L-Tyr to L-DOPA, the rate-limiting reaction in the synthesis of catecholamine neurotransmitters and hormones [1]. Membrane-binding of TH appears to be reduced upon phosphorylation of the enzyme at Ser19, a modification that also regulates its binding to 14-3-3 proteins, a family of ubiquitous proteins.

OBJECTIVES: In this work we aim to understand the molecular determinants and the physiological consequences of membrane-binding for both TH and 14-3-3, separated and when complexed.

METHODS AND RESULTS: As a biological membrane model we are using large unilamellar vesicles (LUVs, also referred to as liposomes) of defined phospholipidic composition. The protein-membrane interactions are investigated by a battery of biophysical methods such as surface plasmon resonance (SPR), circular dichroism (CD), dynamic light scattering (DLS) and leakage of liposome contents by fluorescence. Previous SPR experiments showed that TH_{1-43} (a 43 residue long polypeptide corresponding to the N-terminal region of human TH, isoform 1; hTH1) binds to negatively charged membranes especially in its non-phosphorylated form [2]. Binding of these peptides do not significantly affect the integrity of the membrane, as seen by experiments measuring the leakage of liposome contents. On the other hand, preliminary results show that hTH1 has a high affinity for membranes and a high disruptive activity, while the binding of 14-3-3γ, either alone or complexed with phosphorylated TH_{1-43}, does not affect the integrity of the membrane. Dopamine binding to the enzyme has a protective effect. CD and DLS results show that hTH1 unfolds and aggregates upon membrane-binding.

CONCLUSIONS: Both hTH1 and 14-3-3γ bind to membranes, but only hTH1 seems to disrupt membrane integrity and undergo unfolding and aggregation upon membrane binding. The interaction of hTH1 with dopamine diminishes membrane disruption.

References

[1] Teigen, K., et al. (2007) *Curr Med Chem* **14**, 455−467.

[2] Halskau, O., et al. (2009) *JBC* **284(47)**: 32758−32769.

Dynamic Regulation of Tyrosine Hydroxylase Gene Expression by Key Fate-Determining Transcription Factors during Dopaminergic Neuronal Development in Vivo and in Vitro

Chung-Hyung Kim, Sangmi Chung, Jung-Il Moon, Sunghoi Hong and Kwang-Soo Kim

Molecular Neurobiology Laboratory, Department of Psychiatry and Program in Neuroscience,
McLean Hospital, Harvard Medical School, Belmont, MA, USA

Midbrain dopaminergic neurons are implicated in various neurological and psychiatric diseases such as Parkinson's disease and schizophrenia. Thus, the study of their generation and maintenance is a prerequisite for further understanding of these disease-underlying mechanisms and development of novel therapeutics. In particular, as the first and rate-limiting step of the biosynthesis of dopamine (DA), molecular mechanisms underlying the regulation of tyrosine hydroxylase (TH) gene expression has been the subject of extensive investigations. We have investigated early embryonic development in normal and transgenic mice in which key transcription factors (e.g., Pitx3 and Lmx1a) are inactivated, as a potential model of the pathogenic conditions. In addition, in vitro differentiation system of embryonic stem cells has been used and found to well recapitulate the in vivo differentiation. Furthermore, using TH gene regulation as the functional readout, we have investigated the regulatory cascade and genetic network of extrinsic signaling pathways and intrinsic transcription factors. We found that Wnt1 and Lmx1a form an autoregulatory loop during midbrain DA neuron differentiation of embryonic stem cells, and confirmed its in vivo presence during embryonic development. These studies revealed that TH gene regulation and the phenotypic identity of dopamine neurons are intricately orchestrated by these multiple signals and transcription factors. Furthermore, this novel information can be used to generate a large quantity of functional dopamine neurons that can serve as a platform to study and treat human diseases in which dopamine neurons are degenerated or dysregulated.

Analysis of Tyrosine Hydroxylase Isoforms and Phosphorylation in Parkinson's Disease

Phillip Dickson, Jacqueline Shehadeh, Kay Double, Larisa Bobrovskaya, Stefanie Reyes, Peter Dunkley and Glenda Halliday
School of Biomedical Sciences and Pharmacy, Faculty of Health; University of Newcastle, Callaghan NSW and Neuroscience Research Australia, Randwick, NSW, Australia

Parkinson's disease is a neurodegenerative disease which predominantly targets the neuromelanin containing dopaminergic neurons of the substantia nigra. Unlike other species, tyrosine hydroxylase (TH), the rate-limiting enzyme for dopamine synthesis, occurs in four isoforms (hTH1-4) in the human brain. Using quantitative Western blotting we found hTH1 and hTH2 comprises $\sim$95% of the total amount of TH, while hTH3 and hTH4 represent only $\sim$5% of the total amount. We analysed the distribution of the isoforms in four regions of the human brain; the substantia nigra and ventral tegmental area, the caudate and the putamen in samples from control and Parkinson's disease patients. We found that there was a selective reduction in the level of the hTH1 isoform and a corresponding increase in the level of the hTH2 isoform in samples from patients with Parkinson's disease in all tissues studied. We further showed that there was an increased level of phosphorylation at the serine 40 residue (which is associated with activation of TH) in Parkinson's disease versus control samples in the putamen, caudate and ventral tegmental area. These results may suggest that cells which express a higher proportion of the hTH1 isoform may be more vulnerable to death in Parkinson's disease. This may be related to the fact that the hTH1 isoform is activated to a greater extent in response to stimuli than the hTH2 isoform. The increased level of serine 40 phosphorylation in the remaining dopaminergic neurons in Parkinson's disease patients suggests that TH is activated in these cells in response to the low dopamine levels in the striatum in Parkinson's disease.

Non-dopaminergic Neurons Partly Expressing Dopaminergic Phenotype: Functional Significance and Regulation

Michael V. Ugrumov

Institute of Developmental Biology RAS, Institute of Normal Physiology RAMS, Moscow, Russia

BACKGROUND: Although neurons expressing individual enzymes of dopamine (DA) synthesis, tyrosine hydroxylase (TH) or aromatic L-amino acid decarboxylase (AADC) and lacking DA membrane transporter were discovered two decades ago, functional significance of these monoenzymatic neurons (MN) remained uncertain until recently (Ugrumov, 2009; *J. Chem. Neuroanat.* 38, 241–256).

OBJECTIVES: We have studied the development, distribution, functional significance and regulation of MN.

METHODS AND RESULTS: In vivo and in vitro models, immunocytochemistry, HPLC, ELISA, PCR RT were used. According to our data: (i) MN neurons are widely distributed in the developing brain being more numerous than DA-ergic neurons; (ii) MN expressing complementary enzymes (TH, AADC) produce DA in cooperation, i.e. L-DOPA synthesized in TH neurons is released to the extracellular space and taken up to AADC neurons by the membrane transporter for neutral amino acids followed by DA synthesis; (iii) Although the number of MN decreases in ontogenesis, their fraction in adulthood accounts for 50% of the neurons expressing enzymes of DA synthesis; (iv) TH-expressing MN are in close topographic relations with AADC-expressing MN and bienzymatic neurons suggesting their functional interaction; (v) In adulthood like in ontogenesis, TH-expressing MN synthesize DA in cooperation with AADC-expressing MN and with AADC-containing monoaminergic neurons; (vi) MN serve to compensate DA deficiency under degeneration of tuberoinfundibular DA-ergic neurons; (vii) Neurotoxin-induced degeneration of nigrostriatal DA-ergic neurons provokes an increase of the number of striatal monoenzymatic TH- or AADC-containing nerve fibers and an appearance of MN suggesting a stimulation of DA cooperative synthesis; (viii) Catecholamines inhibit TH expression in non-dopaminergic neurons; (ix) Serotonin stimulates TH expression in tuberoinfundibular neurons.

CONCLUSIONS: (i) Numerous MN are widely distributed in the brain in ontogenesis and adulthood; (ii) MN expressing complementary enzymes (TH, AADC) produce DA in cooperation; (iii) Cooperative DA synthesis is a compensatory reaction in neurodegenerative diseases.

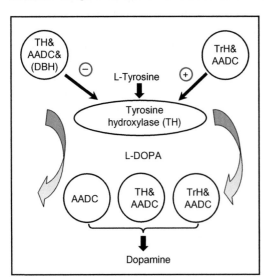

DBH, DA-β-hydroxylase; TrH, tryptophan hydroxylase; other abbreviations (see text).

Developing Brain as an Endocrine Organ: Catecholamines Secretion and Endocrine Action

Julia Saifetyarova, Anna Sapronova and Michael Ugrumov

Institute of Developmental Biology RAS, Institute of Normal Physiology RAMS, Moscow, Russia

BACKGROUND: We tested our hypothesis that the brain operates as an endocrine organ over the ontogenetic period from the neuron origin to the establishment of the blood-brain barrier (BBB) (Ugrumov, 2010; Neurochem. Res. 35, 837−850) using neuron-derived dopamine (DA) and noradrenaline (NA) as markers of the brain endocrine function.

OBJECTIVES: To prove that before the BBB establishment brain-derived catecholamines are delivered from the developing brain to general circulation providing endocrine action on peripheral targets.

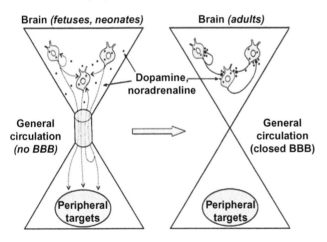

METHODS AND RESULTS: The DA concentration in plasma of rats on the 18th embryonic day (E18), E21, the 3rd postnatal day (P3) was sufficient for providing endocrine action on peripheral targets whereas it dropped to almost undetectable level at P30 after the BBB establishment suggesting that DA circulating in perinatal rats is secreted by the brain. The selective inhibition of DA synthesis with α-methyl-p-tyrosine in the rat brain on P3 resulted in a drop of the DA concentration in plasma. There was no correlation between the ontogenetic evolution of circulating NA and the BBB development as NA is secreted by the adrenals in addition to the brain. Nevertheless, the delivery of the brain-derived NA to the general circulation before the BBB development was proved by the observation of the drop of the NA concentration in plasma on P3 following the α-methyl-p-tyrosine-induced inhibition of NA synthesis. It has been demonstrated *in vitro* that DA in the same concentration as in peripheral blood of perinatal rats inhibited prolactin secretion from the pituitary under its perfusion with Krebs-Ringer solution whereas haloperidol stimulated the prolactin secretion from the pituitary incubated in plasma of rats on P3.

CONCLUSIONS: Brain-derived catecholamines are delivered to the general circulation in ontogenesis before the BBB development providing an endocrine action on peripheral targets that is in line with our hypothesis about the developing brain as an endocrine organ.

Brainstem DOPAergic System

Yoshio Goshima

Yokohama City University School of Medicine, Japan

L-3,4-Dihydroxyphenylalanine (DOPA) has been recognized as merely an intermediate in the catecholamine biosynthetic pathway. Since 1986, we have proposed that DOPA plays a role of neurotransmitter as well (Goshima et al., Brit J Pharmacol 89: 229, 1986; TiPS 14: 119, 1993). There are neurons that are tyrosine hydroxylase (TH) positive but aromatic amino acid decarboxylase (AADC)-negative, suggesting that these neurons probably contain DOPA as an end product. DOPA is released in a transmitter-like manner in vitro and in vivo (Goshima et al., J Neurochem 50, 1725, 1988; Yue et al., Neuroscience 62, 145, 1994). Under the inhibition of AADC, DOPA can produce pre- and postsynaptic responses. For example, microinjection of DOPA into the nucleus tractus solitarii (NTS), induces depressor and bradycardic responses in the anaesthetized in anaesthetized rats. This action is antagonized by DOPA cyclohexyl ester (DOPA CHE), a competitive antagonist of DOPA (Jpn J Pharmacol, 75, 305, 1995). However, specific receptor(s) for DOPA have not been identified yet. Recently, the gene product of ocular albinism 1 (OA1) was shown to possess DOPA-binding activity. Herein we investigate whether OA1 is a functional receptor of DOPA in the NTS. The effects of DOPA microinjected into depressor sites of the NTS on blood pressure and heart rates were examined in anesthetized rats. DOPA CHE microinjected into the NTS blocked phenylephrine-induced bradycardic response. The specific binding of [3H]-DOPA in OA-1 expressing cell line was displaced by DOPA CHE with a Ki value of 14.1 mM. Immunohistochemical examination revealed that OA1-positive cells adjacent to tyrosine hydroxylase (TH)-positive cell bodies and nerve fibers were detected in depressor sites of the NTS. In animals treated with the OA1 shRNA, OA1 expression in the NTS was suppressed, and depressor and bradycardic responses to DOPA microinjected into the NTS were almost completely blocked, while those to glutamate were not affected. We propose that OA1 is a receptor for DOPA that mediates baroreceptor reflex in the NTS. Ourfinding will provide insights into new physiological roles of DOPA, and may lead to a fundamental review of levodopa therapy.

Imaging Norepinephrine Transporters in Humans; Translational Research with PET

Yu-Shin Ding

Center for Biomedical Imaging, New York University School of Medicine, New York, NY, USA

The involvement of NET in the pathophysiology and treatment of ADHD, substance abuse, neuro-degenerative disorders and depression has long been recognized; however, many of these important findings have resulted only from studies in vitro using post-mortem tissues. Thus far, these results have never been verified via in vivo methods because brain imaging of NET in living systems has been hampered due to the lack of suitable radioligands. The fact that all three monoamine transporters (DAT, NET and SERT) are involved in various neurological and psychiatric diseases places an increased sense of urgency on the development of suitable and specific NET ligands so that we will be able to tease out the roles of individual transporters underlying specific CNS disorders. For example, cocaine binds to DAT, NET and SERT with comparative affinity, and its effect on DAT has been well characterized; however, its effect on NET has not been studied due to the lack of specific NET ligands. Furthermore, drugs such as methylphenidate (MP), amphetamine, desipramine and atomoxetine, used for the treatment of ADHD and depression, are not selective and bind NET with high affinities, although this has never been shown in vivo. It is therefore of importance to develop specific NET ligand to assess the NET occupany of these CNS drugs in order to better understand their therapeutic mechanisms. We have recently developed several PET tracers to image NET, and of these (S,S)-[11C]MRB ([11C]MRB) and its derivatives show specific localization and highly encouraging binding kinetics in living systems with PET and have proven to be by far the most promising NET tracers (Ding et al., 2005−2012). Our translational research using [11C]MRB has shown the up-regulation of NET in cocaine dependence. We have also demonstrated a significant age effect of NET in humans. The ability to measure NET occupancy of atomoxetine in humans and non-human primates provides a better understanding of its therapeutic effects in ADHD and depression. Our study in humans showed that oral MP occupies NET in a dose-dependent manner, and that oral MP almost completely occupies NET at clinically relevant doses, with an even more potent ED50 (in vivo) than that for DAT, strongly suggesting the role of NET in ADHD. Our recent study of brown adipose tissue (brown fat, BAT) showed that BAT can be specifically labeled with [11C]MRB, supporting a NET-PET strategy for imaging BAT in humans to study obesity.

Imaging the Vesicular Monoamine Transporter (VMAT2) in Neurodegenerative Diseases

Michael Kilbourn

Department of Radiology, University of Michigan Medical School, Ann Arbor, MI 48109, USA

The vesicular monoamine transporter type 2 (VMAT2) is a protein specifically found in the presynaptic vesicles of all monoaminergic neurons, where it functions to transport newly synthesized or recovered neurotransmitters into the lumen of the vesicle. As one of the monoamine-neuron specific biochemical functions (together with the enzymes necessary for biosynthesis of monoamines, the neuronal membrane transporters responsible for reuptake of synaptic monoamines, and perhaps presynaptic monoamine receptors) the VMAT2 has been targeted for development of high-affinity radioligands useful for non-invasive in vivo imaging (Positron Emission Tomography, PET, or Single Photon Emission Computed Tomography, SPECT) of monoaminergic neuronal densities in human health and disease. Successful radioligands labeled with carbon-11 (t1/2 = 20 min) or fluorine-18 (t1/2 = 110 min) have been synthesized, all based on the structure of tetrabenazine (TBZ), a high affinity VMAT2 ligand developed in the mid-20thcentury. This radioligand development effort culminated in optimized radioligands (+)-a- [11C]dihydrotetrabenazine ((+)-a -[11C] DTBZ) and (+)-a -9-O-(3-[18F]fluoropropyl)dihydrotetrabenazine ((+)-a -[18F [FP-DTBZ); the radioligands are single stereoisomers (2R,3R,11bR absolute configuration) and represented work that was the first to separate and identify the stereoisomers of tetrabenazine-derived molecules. The VMAT2 radioligands have now been studied in a wide variety of neurodegenerative diseases in human subjects (Parkinson's disease, Parkinson's disease with dementia, Alzheimer's disease, Lewy body disease, Progressive Supranuclear Palsy, Multiple Sysyem Atrophy, Corticobasal Degeneration). The most studied condition has been Parkinson's disease, where losses of in vivo radioligand binding have been consistently observed in the striatum, with the regional changes (unilateral presentation early, and putamen > caudate) typical for losses of dopaminergic innervation. Changes of in vivo VMAT2 radioligand binding in other degenerative neurological diseases may be different and can be valuable in assigning correct diagnoses. For example, subjects with dementia with Lewy bodies (DLB) show bilateral losses of VMAT2 radioligand binding throughout the striatum, a pattern clearly different from subjects with dementia due to Alzheimer's disease where the striatal dopaminergic innervation remains essentially intact. VMAT2 radioligand imaging is thus complementary to studies utilizing of 6-[18F]fluoroDOPA (marker of dopamine synthesis) and radioligands for the neuronal membrane dopamine transporter (DAT). These three radiotracer imaging methods delineate different aspects of the dopamine terminals and thus may be subject to different disease- or drug-modifying regulatory effects, not all of which have been fully investigated. The ability of VMAT2 radioligand imaging to correctly identify losses of monoaminergic innervation in numerous neurological diseases is however clearly established.

GTP Cyclohydrolase Regulation: Implications for Brain Development and Function

Hiroshi Ichinose

Graduate School of Bioscience and Biotechnology, Tokyo Institute of Technology, Yokohama, Japan

BACKGROUND: Tetrahydrobiopterin (BH4) is essential for biosynthesis of dopamine, noradrenaline, and serotonin, as a cofactor for tyrosine hydroxylase (TH) and tryptophan hydroxylase (TPH). Co-ordinated action of the monoamine neurotransmitters is pivotal for normal brain development and function. GTP cyclohydrolase (GCH) is the first and rate-limiting enzyme for BH4 biosynthesis. Genetic defect in an allele of the GCH gene can be a cause of dopa-responsive dystonia due to partial BH4 deficiency.

OBJECTIVES: In order to explore transcriptional control of the GCH gene, we analyzed the signaling pathway and the cis-acting elements in PC12D cells and RAW264 cells. Next, we produced two-kinds of BH4-deficient transgenic mice, genetically rescued-*Pts*KO and *Spr*-KO mice, to reveal physiological relevance of BH4 in the brain. *Pts* and *Spr* are the second and third enzymes for BH4-biosynthesis, respectively.

METHODS AND RESULTS: Induction of the GCH gene by NGF in PC12D cells was blocked by the treatment of U0126, a MEK1/2 inhibitor. LPS greatly enhanced the expression of GCH in RAW264 cells, and it was partly suppressed by U0126 or by TLCK due to inhibition of NF-κB pathway.

Immunohistochemical analyses of genetically rescued-*Pts*KO mice showed that TH-immunoreactivity in the lateral region of the striatum was weaker than that in the medial region or in the nucleus accumbens, as observed in the postmortem brains of patients with Parkinson's disease. We also found that motor coordination of limbs was disorganized in genetically rescued-*Pts*KO mice judging from a beam-walking test and a tail-suspension test, whereas *Spr*-KO mice developed a parkinsonian phenotypes like brady-kinesia and tremor.

CONCLUSIONS: Our data suggest physiological relevance of BH4 metabolism in brain development and pathophysiology of neuropsychiatric disorders.

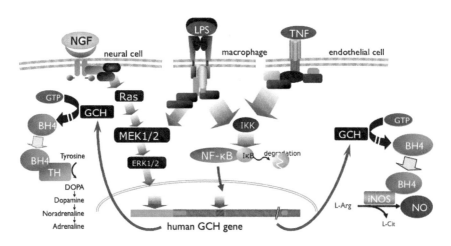

Neonatal Diagnosis of Menkes Disease by a Pattern of Plasma Catechols

CS. Holmes[1], DS. Goldstein[1] and SG. Kaler[2]

[1]Clinical Neurocardiology Section, CNP/DIR/NINDS; [2]Molecular Medicine Program, NICHD,
National Institutes of Health, Bethesda, MD 20892-1620, USA

BACKGROUND: Menkes disease is a lethal X-linked recessive neurodegenerative disorder resulting from mutation of a copper-transporting ATPase, ATP7A. Early treatment with copper injections may prevent death and illness, but presymptomatic detection has been hindered by the inadequate sensitivity and specificity of diagnostic tests. Biochemical features of Menkes disease include relatively high concentrations of dopamine (DA) and its metabolites compared to norepinephrine (NE) and its metabolites, presumably because dopamine-beta-hydroxylase (DBH) requires copper as a co-factor. We explored the relative diagnostic efficiencies of levels of catechol analytes, alone or in combination, in a cohort of neonates at genetic risk for Menkes disease.

METHODS: Plasma from at-risk infants less than 30 days old (n = 44) were assayed prospectively for DA, NE, and other neurochemicals in the catecholamine biosynthetic pathway. Of the 44, 25 were diagnosed subsequently on clinical, biochemical, or molecular grounds as having Menkes disease.

RESULTS: Compared to at-risk infants who did not develop the Menkes disease phenotype, those later shown to have Menkes disease had high plasma DA ($p < 10^{-6}$) and low NE ($p < 10^{-6}$) levels. Considered alone, neither DA nor NE levels had perfect sensitivity, whereas the ratio of DA:NE was higher in all affected than in all unaffected subjects. Analogously, levels of the DA metabolite, dihydroxyphenylacetic acid (DOPAC), and the NE metabolite, dihydroxyphenylglycol (DHPG), were imperfectly sensitive, whereas the DOPAC:DHPG ratio was higher in all affected than in all unaffected subjects.

CONCLUSIONS: Plasma DA:NE and DOPAC:DHPG ratios are remarkably sensitive and specific for diagnosing Menkes disease in at-risk newborns, enabling prompt institution of treatment.

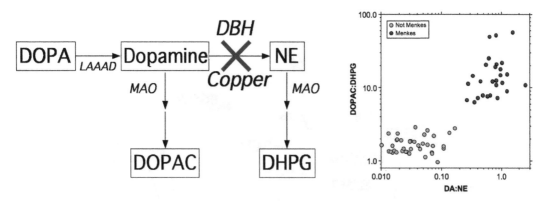

(Left) Catechol metabolic pathways proximal and distal to DBH. Decreased DBH activity in Menkes disease would be expected to increase catechol levels proximal to DBH and decrease levels distal to DBH. (Right) Individual values for DOPAC:DHPG and DA:NE ratios in newborns at risk for Menkes disease. Green circles: unaffected; red circles: affected. Note elevated DOPAC:DHPG and DA:NE ratios in all affected and none of the unaffected at-risk newborns.

Catecholamine Metabolites Affected by the Copper-Dependent Enzyme Dopamine-Beta-Hydroxylase Provide Sensitive Blood and Csf Biomarkers for Viral Gene Therapy in Menkes Disease

Stephen G. Kaler, M.D.

Molecular Medicine Program, NICHD, NIH

BACKGROUND: Menkes disease is a lethal infantile neurodegenerative disorder of copper metabolism caused by mutations in a P-type ATPase, ATP7A. Current treatments are limited and thus we are exploring the potential of innovative therapeutic approaches for this illness. Among its multiple cellular tasks, ATP7A transfers copper to the lumina of the Golgi network and secretory granules, where the copper-dependent enzyme dopamine-beta-hydroxylase (DBH) converts dopamine to norepinephrine. In an animal model that recapitulates the Menkes disease phenotype, the *mottled-brindled* (*mo-br*) mouse, we evaluated the value of DBH-related neurochemical metabolites as biomarkers for response to novel treatment approaches.

OBJECTIVES: We sought to evaluate two specific approaches, adeno-associated viral-mediated ATP7A gene addition and L-threo-dihydroxyphenylserine (L-DOPS) to correct brain neurochemical abnormalities in a mouse model of Menkes disease, and assess the role of DBH deficiency in the pathophysiology of this disease.

METHODS AND RESULTS: We treated *mottled-brindled* (*mo-br*) mice, using combination brain-directed AAV5 gene transfer plus copper or systemic L-DOPS, and measured the concentrations of brain catechols influenced by DBH activity by high performance liquid chromatography. Both treatments improved DBH activity but only gene therapy correlated with increased life span and decreased neuropathology. We conclude that L-DOPS alone does not correct the fundamental defect in copper transport responsible for neurodegeneration in Menkes disease and that partial restoration of ATP7A copper transport function is needed for optimal outcomes. Nonetheless, L-DOPS may ameliorate symptoms of noradrenergic hypofunction in some Menkes disease survivors and be useful, in combination with other treatment. Correction of deficient activity of peptidylglycine alpha-amidating mono-oxygenase (PAM) by improved copper availability may be important in the gene therapy rescue of *mo-br* mice.

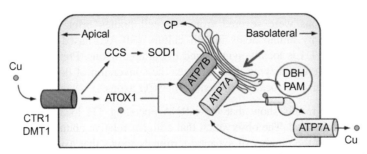

Modified from Kaler et al. Nat Rev Neurol 2011.

Are the Enzymes of the Catecholamine Biosynthetic Pathway Locally Synthesized in the Axon?

Noreen M. Gervasi[1], Sanah N. Vohra[1], Margaret A. MacGibeny[1], Amar N. Kar[1], Anthony E. Gioio[1,2], Anthony J. Makusky[2], Renee L. Olano[2], Sandford P. Markey[2] and Barry B. Kaplan[1]

[1]NIMH-LMB, Bethesda, Maryland, United States; [2]NIMH-LNT, Bethesda, Maryland, United States

BACKGROUND: Tyrosine hydroxylase (TH) is the enzyme that catalyzes the rate-limiting step of catecholamine synthesis, the conversion of tyrosine to L-DOPA. In noradrenergic neurons, TH has been shown to be synthesized in the cell soma and shipped to the axon through anterograde axonal transport. However, studies in recent decades have shown that axons contain a heterogeneous population of mRNAs that can be locally translated into proteins, raising the possibility that TH might be locally synthesized in presynaptic terminals.

OBJECTIVES: To test the hypothesis that TH is locally synthesized in axons.

METHODS AND RESULTS: To isolate pure axonal mRNA and protein, superior cervical ganglion (SCG) neurons from 3-day old rats were plated in the central compartment of Campenot chambers. After 7-9 days in culture, axons were harvested from the side compartments, devoid of cell somas and glial cells. RT-PCR and fluorescence in situ hybridization (FISH) analyses showed that TH mRNA is present in axons. Interestingly, a significant increase in the relative abundance of TH mRNA was observed when axons were grown under culture conditions that led to an increase in the number of synaptic terminals. In order to label locally synthesized proteins, bio-orthogonal non-canonical amino acid tagging (BONCAT) methodology was used. The axonal compartments of Campenot chambers were incubated for 4-6 h with medium containing the methionine analog L-azidohomoalanine (AHA). Newly synthesized AHA-labeled proteins were cross-linked to biotin and affinity purified by incubation with streptavidin-conjugated beads. Protein extracts were then analyzed by Western blotting. AHA-labeled TH was detected in protein extracts of axons incubated with AHA but not in the corresponding neuronal somas, indicating that AHA-labeled TH was specifically synthesized in axons. Furthermore, AHA-labeled TH was detected in axonal protein extracts even when axons were severed from the cell bodies prior to incubation with AHA. Stable isotope labeling by amino acids in cell culture (SILAC)-based mass spectrometric (MS) analysis confirmed that TH is locally synthesized in cultured axons. Preliminary data on dopamine beta hydroxylase (DBH), the enzyme that catalyzes the conversion of dopamine to norepinephrine, suggest that DBH may also be locally synthesized in SCG axons.

CONCLUSIONS: Our results show that the mRNA encoding TH localizes to the axon and that TH can be locally synthesized. The observation that a higher relative abundance of TH mRNA correlated with an increase in the number of synaptic terminals shows that axonal TH mRNA expression is inducible, and suggests that it may be regulated during neuronal differentiation. Taken together, these results show that axonal TH mRNA is functional. Finally, preliminary data on DBH point to the possibility that multiple components involved in catecholamine synthesis are locally synthesized in axons.

Effects of Missense Mutations in Tyrosine Hydroxylase (TH) Found in Patients with Neurological Disorders Attributed to TH Deficiency

Agnete Fossbakk, Per M. Knappskog, Aurora Martinez and Jan Haavik

Departments of Biomedicine, and Clinical Medicine, University of Bergen, Norway

BACKGROUND: Congenital tyrosine hydroxylase deficiency (THD) is found in DOPA responsive dystonia and other related neurological syndromes with predominantly motor symptoms. From 1995–2012, approx. 50 different disease related missense mutations in the exons of TH gene have been reported. As treatment of THD and Parkinson's disease with L-DOPA or dopamine agonists often is inadequate and may only provide temporary symptom relief, searches for treatments that directly target the malfunctioning TH are being conducted.

OBJECTIVES: To perform a systematic genotype/phenotype analysis of TH mutations in THD and to investigate the pathogenic mechanisms of mutated TH found in patients with THD.

METHOD AND RESULTS: All human TH mutations with available clinical phenotype data, not previously studied, were produced using isoform 1 of human TH (TH1) as template (Figure). TH was expressed in E. coli and purified on a heparin-Sepharose column. For 13 out of 23 TH variants, the yield of soluble enzyme was less than 50% compared to wild type (wt) TH. In vitro activity measurements performed at saturating concentrations of substrates showed that the mutated forms of TH had specific activities ranging from 0–65% compared to wt-TH. For the mutants with high residual activity, detailed characterization of kinetic values revealed large variations in Vmax and Km for tyrosine, tetrahydrobiopterin and O2 (range: 10–400% of wt-TH).

INTERPRETATION AND CONCLUSIONS: The great heterogeneity in solubility, stability and enzymatic activity of the mutated forms of TH, indicates that different pathogenic mechanisms may be involved in the neurological syndromes related to THD. Detailed kinetic studies are in progress to gain further insight in the reaction mechanisms of mutated enzyme, to better understand the pathogenic effect of mutations in TH and to find stabilizing compounds that may reverse the negative effects of the mutations on the enzyme activity.

Tetrahydrobiopterin Deficiency Impairs Postnatal Increase of TH Protein via Insufficient Dopamine Biosynthesis

Daigo Homma[1], Junko P. Kondo[1], Setsuko Katoh[2] and Hiroshi Ichinose[1]

[1]*Tokyo Institute of Technology, Yokohama, Kanagawa, Japan;* [2]*Meikai University, Sakado, Saitama, Japan*

BACKGROUND: TH requires tetrahydrobiopterin (BH4) as the essential cofactor. A genetic defect in a BH4-biosynthetic enzyme causes Dopa-responsive dystonia (DRD), DRD is a child-onset dystonia characterized by the decrease of dopamine and the loss of tyrosine hydroxylase (TH) protein in the brain, while the molecular mechanism for its developmental onset and progress is poorly understood.

OBJECTIVES: This study aims to investigate the effect of *sepiapterin reductase* (*Spr*) deficiency on BH4 availability and on the development of dopaminergic neuron during the postnatal period. Spr catalyzes the final step of BH4 biosynthesis.

METHODS AND RESULTS: The BH4 content in the brain of *Spr* KO neonate was significantly lower (ca. 26%) than that of wildtype and there was no significant age-dependent change in each genotype until postnatal day 14 (P14). The contents of dopamine and TH protein in *Spr* KO mouse were not significantly affected in neonatal period, whereas their developmental increases were totally suppressed in *Spr* KO mouse. No obvious abnormality in the morphology of dopaminergic neuron was observed. Repeated administration of BH4 or dopa from P7 to P13 twice a day restored not only dopamine, but also TH protein level in the brain of *Spr* KO mouse.

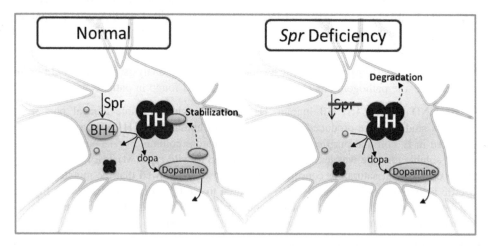

CONCLUSIONS: Genetic deletion of *Spr* causes moderate BH4 deficiency, although its impact on dopaminergic system is limited at neonatal period. The effect of such moderate BH4 deficiency on dopamine and TH protein levels becomes prominent with postnatal development. The therapeutic effect of BH4 and dopa on TH protein level we observed indicate that synergistic interplay between dopamine and TH is crucial for the postnatal development of dopaminergic system, and BH4 deficiency impairs this process via insufficient biosynthesis of dopamine. Our data suggest that the developmental impairment of dopaminergic system may underlie the pathophysiological mechanism for the onset and progress of DRD.

Selective Ablation of Dopamine Beta-Hydroxylase Neurons (Subpopulation of TH Neurons) in the Brain: New Insights into Brain Catecholaminergic Circuitry and Catecholamine-Related Diseases

Keiichi Itoi[1,2], Naoya Sugimoto[1], Katsuya Uchida[1], Toshimitsu Fuse[1], Gopal Das[1], Shinji Ohara[3] and Kazuto Kobayashi[4]

[1]Lab Info Biol, Grad Sch Info Sci; [2]Dept Neuroendocrinol, Grad Sch Med, Tohoku Univ, Sendai, Japan; [3]Dept Neurol, Matsumoto Med Cent, Matsumoto, Japan; [4]Dept Mol Genet, Fukushima Med Univ, Fukushima, Japan

BACKGROUND: The locus ceruleus is the largest noradrenergic (NA) nucleus in the brain. It has been implicated in consciousness, alertness, wakefulness, cognition, anxiety, and depression. Degenerative loss of LC-NA neurons, in addition to nigrostriatal dopaminergic neurons, is marked in Parkinson's disease, but it is not clear whether the defective NA system is related to part of the diverse motor and non-motor symptoms of Parkinson's disease. Neurotoxins, including DSP-4 and 6-OHDA, have been used in previous studies for ablating the LC-NA neurons selectively. However, there are disadvantages in the use of the neurotoxins: these include the variability in thoroughness of destruction among animals and selectivity of the toxin for subpopulation of catecholaminergic neurons. In addition, regeneration of neurons follows in weeks or months after administration of toxins. Dopamine-beta hydroxylase knockout mouse is a useful tool, but all NA neurons are disabled from the embryonic stages.

OBJECTIVES: To overcome these difficulties, we employed the immunotoxin-mediated neuronal targeting for ablation of NA neurons in a subpopulation-selective manner.

METHODS AND RESULTS: Transgenic mice were used in which human interleukin-2 receptor alpha subunit (Tac) is expressed under the promoter of dopamine beta-hydroxylase. The recombinant immunotoxin, composed of the Fv fragment of an anti-Tac monoclonal antibody fused to a truncated form of Pseudomonas exotoxin [anti-Tac(Fv)-PE38], was injected bilaterally into the LC of the transgenic mouse. The LC-NA neurons disappeared almost completely, and tissue noradrenaline was depleted in brain regions that receive NA inputs from the LC. Mice treated with the immunotoxin presented increased anxiety-like behaviors, increased depression-like behaviors, and increased sensitivity to thermo-nociceptive sensation.

CONCLUSIONS: Thus, the LC may be related to part of the pathogenesis of anxiety disorders, depression, and/or pain, which are known to be encountered occasionally in patients with Parkinson's disease.

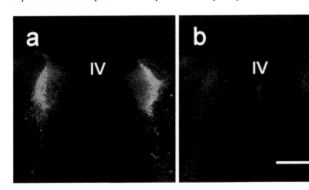

a. The LC-NA neurons in a control (sham-operated) mouse. b. By 10 days following injection of the immunotoxin into bilateral LC, the TH-immunoreactive LC-NA neurons disappeared completely. IV, 4th ventricle; scale bar = 1mm.

Neural Circuit Mechanism for Learning Dependent on Dopamine Transmission: Roles of Striatal Direct and Indirect Pathways in Sensory Discrimination

Kazuto Kobayashi, Ryoji Fikabori and Kayo Nishizawa

Department of Molecular Genetics, Institute for Biomedical Sciences, Fukushima Medical University School of Medicine, Fukushima, Japan

BACKGROUND: The dorsal striatum plays an important role in learning processes contributing to instrumental motor actions. It receives excitatory inputs from many cortical areas and the thalamic nuclei and projects to the output nuclei through two major pathways-so-called direct and indirect. The balance between opposing inputs from these two pathways is considered to be critical in motor control, through regulating basal ganglia output activity. However, how these neural pathways control the learning processes of instrumental actions is mechanistically unclear.

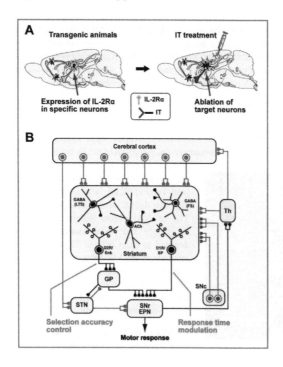

OBJECTIVES: These studies define the behavioral roles of striatal direct and indirect pathways in the performance of sensory discrimination by using immunotoxin (IT)-mediated cell targeting, which eliminates specific neuronal types from a complex neural circuitry in the brain.

METHODS AND RESULTS: IT targeting of striatal D_1 receptor-positive neurons in mice resulted in a moderate level of elimination of the direct pathway. Targeting of the pathway from the whole region of the dorsal striatum lengthened the response time but did not affect the accuracy of response selection in visual discrimination. In addition, subregion-specific pathway targeting revealed that the delay in learned motor response was generated by the elimination of the direct pathway arising from the dorsomedial striatum (DMS) but not from the dorsolateral striatum (DLS). These findings indicate that the direct pathway, in particular from the DMS, contributes to the regulation of response time in the execution of visual discrimination. In addition, IT targeting of the striatal neuronal type containing dopamine D_2 receptor in the DLS of transgenic rats resulted in selective, efficient elimination of the indirect pathway. This elimination impaired the accuracy of response selection in auditory conditional discrimination. These results demonstrate that the DLS-derived indirect pathway plays an essential role in the execution of conditional discrimination, showing its contribution to the control of selection accuracy of learned motor responses.

CONCLUSIONS: Our results suggest that the striatal direct and indirect pathways act to cooperatively regulate the accuracy and response time of learned motor actions in the performance of discrimination learning.

Regulation of Tyrosine Hydroxylase by Ser19-Phosphorylation-Dependent Binding to 14-3-3γ

A. Jorge-Finnigan, M. Ying and A. Martinez

Dept. of Biomedicine, University of Bergen, Norway

INTRODUCTION: Tyrosine hydroxylase (TH) is the catecholamine synthesis rate limiting enzyme and a phenotypic hallmark of several neurodegenerative disorders such as tyrosine hydroxylase deficiency or Parkinson's disease, where it is decreased.

GOALS: Study TH regulation and stability taking into account its cellular distribution and its relation with 14-3-3 proteins, notably γ-isoform, as well as with other partner proteins and membranes.

RESULTS: In PC12 cell line, endogenous phosphorylated Ser19 TH (Ser19P TH) is present in the cytosol and nucleus; Ser31P TH localizes near the nucleus, while Ser40P TH is distributed in the cytoplasm and cell prolongations. 14-3-3γ signal is present in the cytoplasm as well as in the nucleus, and can be found in some of the areas where THS19P is also present (see Figure). Over-expression of GFP-TH in HEK293 cells followed by immnunodetection of its phosphorylated forms revealed a weak signal for THS31P while THSer40 can be phosphorylated as THS40P and GFP signals overlap (see Figure). THS19P results are currently under careful study. Knock-down of endogenous 14-3-3γ in PC12 cells by siRNA suggests that, in the absence of 14-3-3γ, THS19P signal decreases considerably, indicating that 14-3-3 could play a role in TH stability and distribution.

CONCLUSIONS: A preferential cellular distribution of each phosphorylated TH form is observed, which could be dependent on partner proteins, such as 14-3-3 in the case of Ser19P TH. Additional knowledge regarding the mechanisms underlying this observation will be key to understand TH function and regulation in health and in disease.

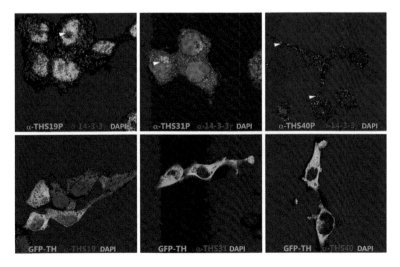

Immunodetection of TH phosphorylated forms. Upper row shows co-detection of THS19P, THS31P or THS40P with 14-3-3γ in PC12 cell. Lower row shows detection of phosphorylated Ser19, Ser31 and Ser40 in GFP-TH fusion protein in HEK293 cells.

Modeling the Dynamics of Dopamine Biosynthesis and its Regulation by Tyrosine Hydroxylase

Rune Kleppe[1,3], Sadaf Ghorbani[1], Agnete Fossbakk[1], Aurora Martinez[1,2] and Jan Haavik[1,2,3]

[1]*Department of Biomedicine, University of Bergen;* [2]*KG Jebsen Centre for Research on Neuropsychiatric Disorders;*
[3]*Haukeland University Hospital, Bergen, Norway*

BACKGROUND: Tyrosine hydroxylase (TH) catalyzes the first and rate-limiting step in catecholamine biosynthesis (Fig. 1) and hence represents a key regulatory site. TH activity is regulated by many mechanisms. Here we focus on post-translational regulation of TH by substrate availability, redox regulation, catecholamine feedback inhibition, Ser-phosphorylation and association to 14-3-3 proteins.

OBJECTIVES: Elucidate how different regulatory mechanisms of TH may contribute to the regulation of DA biosynthesis using mathematical modeling.

METHODS AND RESULTS: Reported and experimentally obtained kinetic parameters of TH activity and regulation were incorporated into a quantitative three compartment model involving in total 76 reactions, using the simulation software Copasi. Protein purification, phosphorylation and dephosphorylation were performed as described [1]. Modeling of TH clearly illustrated a bi-phasic activity response to oxygen. Thus, TH operated at sub-saturating conditions at normal brain oxygen levels, but at higher oxygen levels TH activity became limited by increased auto-oxidation of active site Fe(II). This process was amplified by DA binding, low BH$_4$-levels, but inhibited by Ser40 phosphorylation. The effect of 14-3-3 proteins on Ser40 phospho-turnover is not known. We found that Ser19 phosphorylation of TH enabled 14-3-3ζ to act as a gatekeeper on Ser40 phospho-exchange. This was mediated by promoting the rate of phosphorylation over dephosphorylation by 8 fold. Modeling showed that binding of 14-3-3 proteins was necessary to mediate pSer19 induced Ser40 phosphorylation at conditions consistent with basal and stimulated cellular Ser19 and Ser40 phosphorylation stoichiometries. Simulating the situation of an abrupt large release of the vesicular DA pool, suggest that Ser19- and 14-3-3 mediated increase in Ser40 phosphorylation facilitate a more rapid recovery of vesicular DA after its depletion. This is expected to be of greater importance for dopaminergic volume transmission or at conditions with little DA reuptake.

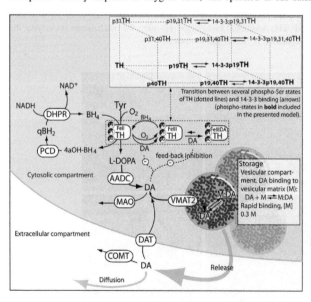

Reference

[1] Toska et al. (2002) *J. Neurochem.* **83**, 775–83.

Reassessment of Intrinsic Dopaminergic Innervation in the Human Enteric Nervous System – Clinical Implications

Eberhard Weihe[1], Martin K.-H. Schäfer[1], Martin Anlauf[2] and Lee Eiden[3]

[1]*Institute of Anatomy & Cell Biology, Philipps University Marburg;* [2]*Institute of Pathology, Heinrich-Heine University Düsseldorf, Germany;* [3]*National Institute of Mental Health, Bethesda, Maryland, USA*

BACKGROUND: Besides extrinsic noradrenergic innervation, there is recent evidence for substantial intrinsic dopaminergic innervation of the human gastrointestinal (GI) tract.

OBJECTIVES: This study aims to differentiate catecholaminergic traits in the intrinsic and extrinsic innervation of the human oesophagus and GI tract, to reveal their proportional relationships with cholinergic and nonadrenergic, noncholinergic (NANC) phenotypes, to determine their target relations and to consider their clinical implications.

METHODS AND RESULTS: For this purpose, immunohistochemistry with antibodies against the vesicular monoamine transporters (VMAT1/2), tyrosine hydroxylase (TH), amino-acid decarboxylase (AADC), dopamine beta-hydroxylase (DBH), the vesicular acetylcholine transporter (VAChT), and vasoactive intestinal peptide (VIP) were used. A substantial subpopulation of intrinsic submucosal and myenteric neurons in the human gut especially in the oesophagus, stomach and duodenum is dopaminergic (TH+/AADC+/DBH-/VMAT2+). The noradrenergic (TH+/AADC+/DBH+/VMAT2+) phenotype is totally absent from intrinsic neurons of either submucosal or myenteric ganglia. Therefore, VMAT2+/TH+/AADC+/DBH+/VMAT2+ nerve fibers represent the extrinsic, postganglionic, noradrenergic sympathetic innervation. The intrinsic innervation of the human oesophagus and gut is divided into three components with respect to cholinergic and catecholaminergic phenotypes: (1) 50–70% cholinergic (VAChT+), (2) up to 20 % dopaminergic (TH+/AADC+/DBH-/VMAT2+) and (3) noncholinergic, nonadrenergic (NANC). Most intrinsic cholinergic neurons code for VIP. Targets of the intrinsic dopaminergic neurons included muscle cells of the circular and longitudinal layers, enteroendocrine cells such as ECL cells (VMAT2+) and EC cells (VMAT1) and mucosa associated lymphoid tissue as well as VMAT2+/TH+ mast cells. Submucosal and myenteric ganglionic plexus as well as blood vessels throughout the gut wall receive noradrenergic innervation.

CONCLUSIONS: We suggest that the intrinsic dopaminergic innervation of human oesophagus and GI tract modulates motility, enteroendocrine secretion and mucosal immune regulation. Dysfunction of intrinsic dopaminergic enteric innervation may be of clinical importance in motility disorders, inflammatory bowel diseases, Parkinson- and dysautonomia-associated syndromes and possibly in neuroimmune regulation of sepsis and cancer.

Dopaminergic innervation of "DOPAergic" mucosal mast cells   **DAergic neurons in myenteric ganglion**

TH   tryptase   merge   TH

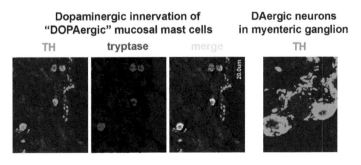

Midbrain Dopamine Neurons are Divided into Different Functional Groups

Masayuki Matsumoto

Primate Research Institute, Kyoto University, Inuyama, Aichi, Japan

Dopamine neurons in the substantia nigra pars compacta (SNc) and the ventral tegmental area (VTA) are key components of the brain's reward system. These neurons are excited by reward and sensory stimuli predicting reward. The excitatory responses increase as the reward value increases. Based on their response property, it has been thought that dopamine neurons encode a value-related signal.

However, we recently showed that not all dopamine neurons encode the value-related signal uniformly. We recorded the activity of dopamine neurons in monkeys during a Pavlovian procedure with appetitive and aversive outcomes (liquid rewards and airpuffs directed at the face, respectively). We found that some dopamine neurons were excited by reward-predicting stimuli and inhibited by airpuff-predicting stimuli, as the value hypothesis predicts. However, a greater number of dopamine neurons were excited by both of these stimuli inconsistent with the hypothesis. These neurons are presumed to encode motivational salience which indicates a quantity that is high for both rewarding and aversive events and is low for motivationally neutral events. Notably, the salience-coding neurons were located more dorsolaterally in the SNc, whereas the value-coding neurons were located more ventromedially, some in the VTA.

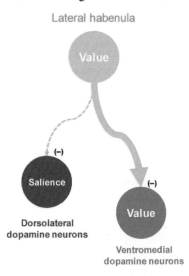

Which brain areas provide dopamine neurons with signals necessary for value- and salience-coding? One major candidate is the lateral habenula which exerts potent negative control over dopamine neurons. We recorded the activity of lateral habenula neurons using the same Pavlovian procedure. We found that these neurons were inhibited by reward-predicting stimuli and excited by airpuff-predicting stimuli, indicating that they encode motivational value. The value signals would then be transmitted to dopamine neurons by inhibiting them. Our findings suggest that the lateral habenula plays an important role as a source of motivational value signal in dopamine neurons, though the source of their motivational salience signal is still unclear.

Cell-, Region- and Species-Specific Expression of VMAT2: Implications for Medical Imaging With High-Affinity Transporter Ligands

Martin K.-H. Schäfer[1], Martin Anlauf[2] and Eberhard Weihe[1]

[1]Institute of Anatomy & Cell Biology, Philipps University Marburg, Institute of Pathology;
[2]Heinrich-Heine University Düsseldorf, Germany

BACKGROUND: Monoamine storage in secretory granules is mediated by the vesicular monoamine transporters 1 and 2 (VMAT1, VMAT2). Analogs of the VMAT2 selective ligand tetrabenazine (TBZ) have been developed as PET tracers for VMAT2 imaging in vivo.

OBJECTIVES: Here we aim to define the various VMAT2 systems with respect to cell type and presence or absence of the various monoamine phenotypes, and their presence in humans and animal models for translational research.

METHODS AND RESULTS: Across species, VMAT2 expression in relation to enzyme markers of the monoaminergic systems revealed the classical dopaminergic, noradrenergic, serotonergic, and histaminergic functional phenotypes of VMAT2+ pathways throughout brain and periphery. 'Non-classical' TH+ neurons lacking VMAT2 exist in distinct regions of rodent and primate brain. We found no evidence for VMAT1 expression in CNS neurons. All sympathetic innervation and enteric dopaminergic innervation is VMAT2-positive. In humans but not in rodents, VMAT2 is contained in coincident cholinergic/noradrenergic sudomotor nerves and in most cholinergic intracardiac neurons. Extra-neuronal VMAT2 expression occurs in adrenomedullary cells, in histaminergic enterochromaffin-like (ECL) cells and in cells of the hematopoietic lineage such as megakaryocytes, platelets, and basophils. VMAT2 is expressed in mast cells, which in humans but not rodents co-express TH. VMAT2 expression occurs in human but not rodent cutaneous Langerhans cells. Pancreatic beta cells of humans and pigs but not of rodents express VMAT2 mRNA and protein.

CONCLUSIONS: This comprehensive characterization of cell- and species-specific VMAT2 expression guides correct interpretion of medical imaging with high affinity PET tracers for VMAT2 throughout the organism both under clinical and translational research conditions. In particular, the utilization of TBZ imaging ligands in estimating loss of beta cell mass is promising in human diabetes and in pig, but not in rodent models of diabetes. *Supported by EU FP7/2007-2013 no. 222980*

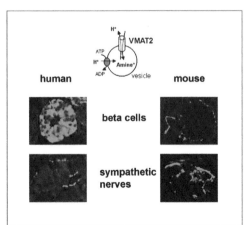

Catecholamine Release and Re-uptake

B

Lee Eiden and David Goldstein

The molecular identification of the proteins responsible for catecholamine re-uptake after physiological release, and acted upon by drugs such as amphetamine and cocaine to elicit their rewarding and addictive properties, were an important advance of the 1990s. The abstracts from the sessions within the Tenth International Catecholamine Symposium convened under the banner of this theme illustrate the important advances made since the beginning of the 21st century.

The plenary session inaugurating this portion of the symposium, by Dr. Mark Wightman, was an important historical reminder that the ability to directly measure catecholamine release in the living brain of experimental animals has positioned the field of neurochemistry to solve fundamental questions such as the mechanisms of reward, addiction and motivation (see also Themes F and G). Another way to pierce the veil between conclusions drawn from molecular biological, biochemical and cell culture experiments, and the control of brain function and behavior by catecholamine neurotransmission, is through the expression in whole animals of genetic variants of the molecules controlling catecholamine release and re-uptake. Thus mutations in NET, DAT, VMAT2 reveal key functions for trafficking domains of these molecules in control of their regulation, including inter-action with the heretofore 'orphan' Trace Amine Associated Receptors (TAARs).

The chromaffin cell was featured in this symposium as the major neuroendocrine depot from which epinephrine is released during both systemic and psychological stress responses. In addition, it is the only existing primary neuroendocrine cell model for studying catecholamine secretion, includ-ing the role of the granin family in secretory vesiculogenesis and regulation of exocytosis. The role of the neuropeptide PACAP as the major effector for epinephrine release from chromaffin cells, and the hormonal role of chromogranin A—and several bioactive fragments of it—upon co-release with epinephrine and norepinephrine from the chromaffin cell was the subject of an entire session in Theme B. It is worth noting that extension of this important work to the mechanisms of release and co-release from sympathetic post-ganglionic neurons has lagged somewhat and represents a clear gap in current understanding of catecholamine release in the peripheral nervous system.

Resonating with Mark Wightman's plenary talk, a session devoted to imaging of dopamine release in vivo featured new connections the actions of drugs of abuse to regiospecific release of DA, as well as an entirely new mechanism of action for amphetamines, based on their ability to affect not only catecholamine transport by DAT, but the trafficking and plasma membrane presentation of the DAT protein.

The final session in this theme was devoted to DA cycling and storage as it pertains to catechol-amine toxicity occurring under conditions of impaired VMAT2 function, chronic pyschostimulant use, and neurodegenerative disease such as Parkinson's (see also Theme F). The intrinsic toxicity of dopamine and other catecholamines unmasked when vesicular storage is compromised is a vivid reminder of Schuldiner's original formulation of the vesicular transporters as members of the TEXAN, or 'toxin extruding antiporter family'.

Detecting Catecholamines – A Journey from Beaker to the Behaving Brain

Mark Wightman

University of North Carolina at Chapel Hill, North Carolina, USA

I began my foray into catecholamines in my postdoctoral research with Ralph Adams at the University of Kansas. Adams had the revolutionary idea that electroanalytical chemistry could be useful for the detection of catecholamines. From this fundamental concept evolved the use of electrochemical detectors with liquid chromatography as well as the use of microelectrodes for in situ sensing of catecholamines. The early research with in vivo electrochemistry was complicated by interference from a variety of substances including catecholamine metabolites and ascorbate. However, a variety of approaches were taken to improve the specificity of the measurements and to adapt them for measurements on a subsecond time scale. Today these approaches are extremely useful to understand the dynamics of catecholamines during release, uptake and signaling. They can be used to probe single exocytotic events from single cells in culture. They also can be used in the brain of awake animals to watch catecholamine signaling during behavior. This approach has been used to confirm dopamine release in response to cues that predict reward as well as to examine norepinephrine signaling in aversive situations.

Functional Disturbances in DA Homeostasis and Signaling Revealed Through the Study of Rare Dopamine Transporter Coding Variants

Randy D. Blakely, Ph.D.

Departments of Pharmacology and Psychiatry, Silvio O. Conte Center for Neuroscience Research, Vanderbilt University School of Medicine, Nashville TN, 37232-8548, USA

Attention-Deficit/Hyperactivity Disorder (ADHD) is the most commonly diagnosed behavioral disorder of childhood, with current estimates ranging from 5-10% of school age children. PET imaging studies and evaluation of common genetic variation, particularly in the dopamine (DA) transporter

(DAT) gene (SLC6A3) and the D4 DA receptor (DRD4) gene, have linked altered DA signaling to risk for ADHD. Additionally, the most commonly prescribed medications for the treatment of ADHD target DAT or DA clearance mediated by the norepinephrine transporter (NET, SLC6A2). In order to move beyond correlation in implicating altered DA signaling in ADHD, we sought evidence for rare, but highly penetrant, gene variation in SLC6A3. To date, we have identified 4 rare, missense DAT coding variants that localize to highly conserved regions of DAT protein. Functional studies, both in vitro and in vivo using transgenic mouse models reveal perturbations induced by these variants in DAT function, trafficking, regulation, and drug responses. Most recently, our studies of the DAT A559V knock-in mouse model provide functional evidence for these animals as a construct-valid ADHD model whose detailed investigation can provide a platform for an integrative analysis of ADHD risk not possible with human subjects.

Norepinephrine Transporter A457P Knockin Mice Display Tachycardia and Anxiety-Like Behavior

Maureen K. Hahn, Jana K. Shirey-Rice, Rebecca Klar, Hugh M. Fentress, Sarah N. Redmon, Tiffany R. Sabb, Jessica J. Krueger, Nathan M. Wallace, Martin Appalsamy, Charlene Finney, Suzanna Lonce, Andre Diedrich and David H. Robertson

Dept. of Medicine, Vanderbilt University School of Medicine Nashville, TN, USA

BACKGROUND: Norepinephrine (NE) serves as a neurotransmitter in both the brain, where it mediates attention, memory emotion, and the response to stress, and in the sympathetic nervous system. NE neurotransmission is involved in the pathogenesis of cognitive and mood disorders and cardiovascular diseases. NE released at synapses in the brain and sympathetic nervous system is actively transported into terminals by the presynaptically-localized norepinephrine transporter (NET), making NET a critical mediator of NE inactivation and presynaptic catecholamine homeostasis. We previously identified a single nucleotide polymorphism in the human NET gene, A457P, in a family with incidence of postural orthostatic tachycardia syndrome, demonstrating highly elevated heart rate and plasma NE upon standing as well as psychiatric symptoms. In vitro expression studies demonstrate that A457P is a loss-of-function transporter with a dominant-negative influence on wild-type (WT) NET.

OBJECTIVES: We generated "knockin" mice expressing NET A457P to determine the contribution of A457P to both psychiatric and cardiovascular phenotypes.

METHODS AND RESULTS: Methods: NE transport and cell-surface biotinylation assays/immunoblotting for NET were performed to determine NET expression and activity. Catecholamines and metabolites were measured by HPLC from plasma and urine samples. Blood pressure radiotelemetry was used to determine blood pressure and heart rate in conscious mice. Mice were exposed to the elevated plus maze and time spent/number of entries in the open versus closed arms and freezing behavior were measured using video analysis (Anymaze). Open field exploration was performed in which total distance traveled and time spent in center versus periphery of the field were calculated. Results: In cortex and hippocampus of A457P heterozygous (HET) knock-in mice, total levels of NET were reduced to 68.9 ± 2.1 and $73.1 \pm 10.6\%$, of WT, and surface levels of NET were reduced to 71.4 ± 5.2 and $80.0 \pm 9.0\%$ of WT ($p < 0.05$-0.01; $N = 4$-6), respectively. Transport levels in A457P HET mice were 60.8 ± 4.8 and $63.3 \pm 5.0\%$ of WT in cortex and hippocampus, respectively ($p < 0.001$, $N = 6$). In the heart and other peripheral, sympathetically-innervated tissues, NE transport was also reduced in A457P HET mice ($p < 0.01$-0.001; $N = 5$). Basal urine DHPG levels were decreased, NE was elevated, and the DHPG to NE ratio was decreased in A457P mice, with similar results in plasma ($p < 0.05$-0.001; $N = 20$). Radiotelemetry recordings demonstrated higher heart rates in A457P mice compared to WT mice (628 ± 15 vs. 590 ± 3 beats per minute; $p < 0.05$; $N = 5$-6). A457P mice spent decreased time in the open arms and increased time in the closed arms of the elevated plus maze (EPM). A457P mice froze more often and for longer durations than WT mice and demonstrated decreased head dips in the EPM, all evidence of anxiety ($p < 0.01$; $N = 16$-18). We also detected anxiety in the open field, where A457P mice spent significantly less time in the center of the field and increased time along the periphery ($p < 0.01$, $N = 16$-17).

CONCLUSIONS: These data support that A457P is trafficking- and activity-deficient in neurons in vivo. These data also demonstrate that A457P is sufficient to recapitulate the neurochemical and cardiovascular phenotypes observed in human carriers. Additionally, A457P mice demonstrate anxiety phenotypes. Notably, patients with tachycardia disorders have a higher incidence of anxiety. Thus, genetic disruption in genes, such as NET, with shared expression in the brain and autonomic nervous system, may explain comorbidity of psychiatric and cardiovascular disease.

In Vivo Mutations in the Trafficking Domain of the Drosophila Vesicular Monoamine Transporter Disrupt Specific Amine-Dependent Behaviors

Anna Grygoruk[1], Audrey Chen[1,2], Ciara R. Martin[1], Niall P. Murphy[1], Nigel T. Maidment[1] and David E. Krantz[1]

[1]*Department of Neurobiology, University of California, Los Angeles, CA 90095;* [2]*Department of Psychiatry and Biobehavioral Sciences and Semel Institute for Neuroscience and Human Behavior, Hatos Center for Neuropharmacology, University of California, Los Angeles, CA 90095, USA*

We are using Drosophila melanogaster as a model system to study the membrane trafficking of the vesicular monoamine transporter (VMAT) in vivo. As in mammals, the Drosophila ortholog of VMAT (DVMAT) is responsible for the vesicular storage and release of all biogenic amines including serotonin, dopamine and octopamine, the fly analog of mammalian norepinephrine. Using newly generated constructs containing an acid-sensitive form of GFP (DVMAT-pHluorins) we first analyzed exo- and endocytosis of wild type and mutant forms of DVMAT at the larval neuromuscular junction. Consistent with the proposed importance of the VMAT carboxy terminus for trafficking in both mammals and flies, we find that simultaneous deletion of both tyrosine-based and dileucine (DiLeu) motifs in this domain, represented by the DVMAT-Δ3 mutant, reduces exocytosis and also decreases the rate of endocytosis at the nerve terminal; mutation of the individual motifs had less dramatic defects suggesting that they may be partially redundant for endocytosis at the synapse. To determine how these mutations would affect localization to synaptic vesicles (SVs), we performed sucrose density fractionation of wild type and mutant forms of DVMAT. We find that deletion of both the tyrosine and DiLeu motifs in DVMAT-Δ3 results in a dramatic deficit in sorting DVMAT to SVs. Surprisingly, deletion of both motifs also led to an increase in the localization of the DVMAT$\Delta\Delta$3 to the cell body, suggesting that the signals that localize proteins to SVs and the nerve terminal may be related or identical. Mutation of the tyrosine motif alone (DVMAT-Y600A) also showed a deficit in sorting to SVs, but was less severe than DVMAT-Δ3. In addition, the Y600A mutant showed increased sedimentation to fractions containing large dense core vesicles (LDCVs), a secretory organelle distinct from SVs and thought to be important for the neuromodulatory effects of most monoamines. To determine the potential behavioral effects of increased amine release from LDCVs or more generally decreased release from both SVs and LDCVs, DVMAT-Y600A, -Δ3 and -wild type were used to genetically rescue a null mutation in the endogenous dVMAT gene. Despite severe trafficking deficits, the DVMAT-Δ3 mutant rescued amine storage and several amine-dependent behaviors indistinguishably from wild type DVMAT. Conversely, DVMAT-Δ3 failed to rescue female fertility, indicating that the circuit underlying this behavior may be more sensitive than others to defects in precise exocytotic release. In addition, rescue with DVMAT-Y600A resulted in a different spectrum of behavioral deficits via a gain of function mechanism, suggesting that shifting the balance of amine release from SVs to LDCVs can have unexpectedly dramatic consequences. These data are the first to demonstrate how altered vesicular transporter trafficking can alter the function of the nervous system as a whole, and provide a powerful system to investigate the mechanisms underlying these effects.

Fate of the Chromaffin Granule Membrane and its Protein Constituents after Fusion

Ronald W. Holz[1], Arun Anantharam[1,2], Daniel Axelrod[1,3] and Mary A. Bittner[1]

[1]*Departments of Pharmacology;* [2]*Department of Biological Sciences, Wayne State University, Detroit, MI;* [3]*Physics and LSA Biophysics, University of Michigan. Ann Arbor, MI*

BACKGROUND: There are multiple potential fates after fusion for the constituents of the granule membrane in neuroendocrine cells. In chromaffin cells the trafficking of the vesicular monoamine transporter (VMAT) and membrane-bound dopamine-β-hydroxylase (DBH), two proteins that help define the catecholamine-secreting phenotype, is inextricably linked to common underlying processes that are critical for the economy of these highly differentiated, secretory cells. Classically it was thought that the granule membrane becomes incorporated into the plasma membrane, with granule constituents being retrieved over minutes through a poorly defined, clathrin-mediated endocytotic pathway. New optical techniques permit more precise elucidation of the fate of granule membrane and its constituents after fusion.

OBJECTIVES: The goal of the following experiments was to more completely determine the pathways taken by the granule membrane and its protein constituents after fusion.

METHODS AND RESULTS: A combination of polarization and TIRFM captured the dynamics of the expanding fusion pore during elevated-K + induced secretion. The method demonstrated that in most cases the granule membrane flattens into the plasma membrane. Surprisingly, this process is regulated by dynamin, the major regulator of membrane fission in endocytosis, probably through dynamin's membrane shaping properties. The pH dependency of the fluorescence of VMAT2-pHluorin revealed that rapid endocytosis occurs following only 10-20% of the fusion events. There is an alternative recycling pathway. Within 20 seconds of fusion, GFP-clathrin light chain associates with exocytotic sites. Immunocytochemistry revealed that DBH, VMAT2 and synaptotagmin remain punctuate and associated for many minutes after fusion. Immunocytochemistry revealed that endogenous clathrin associates with DBH and is detected in DBH-containing endocytic vesicles. Quantitative immunocytochemistry revealed that individual plasma membrane puncta do not retain the full amount of originally deposited DBH and VMAT2. Instead, the amounts in the puncta decline over several minutes, suggesting partial endocytosis of the granule membrane. Indeed, the amounts of DBH and VMAT in individual endocytic vesicles are 10-20% of individual plasma membrane puncta resulting from exocytosis. An inhibitor of dynamin GTPase activity, Dyngo-4A, completely inhibited the process. As previously observed, electron microscopy revealed 100 nm diameter, clathrin- coated vesicles with no evidence of substantially larger vesicles. These results suggest that upon fusion, the chromaffin granule membrane is recycled into the cell by a clathrin- and dynamin-mediated endocytotic 'nibbling' mechanism.

CONCLUSIONS: These studies reveal that the fused granule membrane represents a distinct trafficking domain in the plasma membrane. Its initial curvature is regulated and its ultimate fate determined by two pathways, rapid endocytosis of the entire granule membrane and slower, clathrin- and dynamin-mediated endocytosis. The latter occurs by a 'nibbling' mechanism, consistent with the size mismatch of the chromaffin granule membrane (area corresponding to a 300 nm diameter organelle) and a clathrin-coated vesicle (~100 nm).

Acknowledgements

This work was supported by NIH grant R01-NS38129 to RWH and DA and NIH fellowships T32DA007268 and F32GM086169 to AA. This work benefited from a subsidy for DNA sequencing from University of Michigan Comprehensive Cancer Center.

Catestatin (Human Chromogranin A352-372) Induces Lipolysis and Fatty Acid Oxidation through Regulation of Adrenergic and Leptin Signaling

**Sushil K. Mahata[1,2], Christine U. Vu[1], Stefano Gentile[1,3],
Gautam K. Bandyopadhyay[1] and Daniel T. O'Connor[1,2]**

[1]*VA San Diego Healthcare System San Diego, California, USA;* [2]*University of California, San Diego
La Jolla, California, USA;* [3]*University of Calabria Cosenza Italy*

BACKGROUND: The secretory proprotein Chromogranin A (CHGA in humans, Chga in mice) gives rise to several peptides of biological importance, which include the dysglycemic hormone pancreastatin (PST: CHGA250-301), the vasodilator vasostatin (CHGA1-76), and the antihypertensive, antiadrenergic, cardiosuppressive and angiogenic peptide catestatin (CST: CHGA352-372). The increased adiposity in hyperadrenergic, hyperleptinemic and insulin-sensitive Chga knockout (Chga-KO) mice is due to resistance to catecholamines and leptin.

OBJECTIVES: Since catestatin inhibits catecholamine secretion and catecholamines inhibit leptin secretion, we reasoned that CST would reduce obesity by restoring adrenergic receptor and leptin receptor sensitivity through normalization of catecholamine and leptin levels.

METHODS AND RESULTS: Chronic CST administration (5 μg/g BW/day for 16 days) to Chga-KO mice reduced epididymal fat pad size to WT level and decreased plasma triglyceride levels. This decrease in overall lipid content is caused in part by increased lipolysis as evident by increased glycerol and non-esterified fatty acid levels in plasma. CST treatment of Chga-KO mice also lowered plasma leptin to a level below WT and restored leptin action by reversing the desensitization effect of chronic leptin excess. Since adipose tissue in CST-treated mice showed increased palmitate oxidation but decreased incorporation into lipids, we believe that CST inhibits the expansion of adipose tissue and also promotes fatty acid uptake in liver for oxidation. In liver, CST augmented the expression of acyl-CoA oxidase 1 (Acox1), carnitine palmitoyltransferase 1a (Cpt1a), uncoupling protein 2 (Ucp2), and peroxisome proliferator-activated receptor-alpha (Pparα) genes involved in fatty acid oxidation. In contrast, CST had no effect on the expression of lipogenic genes such as sterol regulatory element-binding protein 1 (Srebp-1) and peroxisome proliferator-activated receptor-gamma (PparΔ. Interestingly, CST stimulated the expression of cluster of zzdifferentiation 36 (Cd36), a transporter mediating cellular uptake of long-chain fatty acids, as well as the lipogenic gene Gpat4 (glycerol-3-phosphate acyltransferase-4). This indicates that CST stimulates fatty acid incorporation into triglycerides but not de novo lipogenesis. Overall, CST appears to promote lipid flux from adipose tissue toward liver for catabolism. CST reversed the hyperleptinemia of Chga-KO mice and improved leptin signaling as determined by phosphorylation of AMPK and Stat3. Both the α-antagonist phentolamine and CST potentiated the lipolytic effects of b-AR agonist iso-proterenol in primary adipocytes. These findings suggest that CST recapitulates the lipolytic effect of the a-AR antagonist phentolamine. This commonality of CST with phentolamine, coupled with its ability to inhibit the action of a-AR agonist phenylephrine, suggests that CST acts by suppressing α-AR signaling.To further clarify the interactions between CST and leptin pathways and to establish CST as an anti-obesity factor, we examined the effects of CST in leptin resistant high fat diet-induced obese (DIO) mice and leptin deficient Ob/Ob mice. CST treatment reduced body weight and adipose mass in DIO mice without reducing food intake. In contrast, CST decreased (by >25%) food intake in Ob/Ob mice. Interestingly, CST enhanced leptin effects on adipose tissue metabolism and signaling in both DIO and leptin-deficient Ob/Ob mice.

CONCLUSIONS: We conclude that CST promotes lipolysis and fatty acid oxidation by blocking α-adrenergic receptor signaling as well as by enhancing leptin receptor signaling. Therefore, we propose CST as a promising candidate for anti-obesity.

Introduction: From Mystery Matrix Proteins to Precise Molecular Actions in Human Disease

Daniel O'Connor

UCSD, USA

By the late 1970s, the chromogranins/secretogranins were known as major structural proteins in chromaffin granules, but function remained a topic for speculation. As a result of sequential revolutions in biotechnology over the past ~30 years (cDNA cloning and sequencing, promoter motif dissection, transgenic and knockout mice, deciphering the scaffold of the human genome, rapid methods to discover human genetic variation, role of non-coding RNAs, such as micro-RNAs in gene expression), variation of chromogranins in human disease has been assigned a more precise and mechanistic role. Here we will discuss genetic involvement of a major chromogranin in a devastating human disease state: hypertensive renal failure (including end stage renal disease requiring lifelong dialysis), especially common in African-Americans. Allelic and haplotype association (case/control) studies established a role for genetic variation in the disease, and followup molecular studies in the 3'-UTR then indicated that the associated variant disrupted a recognition motif for a particular micro-RNA. We will discuss the implications of these findings for diagnosis, pathogenesis, and potentially for novel treatment approaches to the syndrome.

Acknowledgements

Critical participants in these studies, without whom the work would not be possible, include Gen Wen, Bruce A. Hamilton, Fangwen Rao, Yuqing Chen, Kuixing Zhang, Rany M. Salem, Sushil K. Mahata, Manjula Mahata, Stephane Chiron, Jiaur Gayen, Sucheta Vaingankar, Sajalendu Ghosh, Mats Stridsberg, Barry Freedman, Maple M. Fung, Michael G. Ziegler, and Nicholas J. Schork.

A New Take on Uptake: Dopamine Transporters & the Cellular Mechanisms of Amphetamine Action

Susan G. Amara[1,4], **Suzanne M. Underhill**[1,4], **David S. Wheeler**[2], **Conrad W. Hong**[1,4], **Geoffrey H. Murdoch**[3], **Guillermo Romero**[2], **Susan L. Ingram**[5] and **Edda Thiels**[1,4]

[1]*Departments of Neurobiology;* [2]*Departments of Pharmacology and Chemical Biology;* [3]*Departments of Pathology, University of Pittsburgh School of Medicine, USA;* [4]*Departments of Center for Neuroscience at the University of Pittsburgh;* [5]*Department of Neurosurgery, Oregon Health & Science University, USA*

Psychostimulants induce euphoria and hyperactivity by increasing extracellular dopamine. Cocaine potentiates dopamine signaling by acting as non-transported antagonist of the plasma membrane dopamine transporter (DAT). However, the actions of amphetamines are more complex because they also cause the release of dopamine by a non-vesicular mechanism involving substrate exchange and/or reverse transport through DAT and VMAT1, but the precise mechanisms underlying these effects remain unclear. Previous work has shown that amphetamines, unlike cocaine and other non-transported blockers, have the ability to stimulate internalization of DAT from the cell surface. Although amphetamines and cocaine both induce euphoria, arousal, and hyperlocomotion, amphetamine exposure has behavioral, neuroadaptive, and neurotoxic consequences not associated with cocaine use. To identify a potential molecular basis for the unique actions of amphetamines, we undertook a detailed study of how they stimulate DAT internalization. We found that amphetamine (AMPH) acts within the cytoplasm to activate Rho GTPases, triggering DAT internalization through a Rho- and dynamin-dependent pathway. Cytoplasmic AMPH also stimulates cAMP production, which leads to Rho inactivation by PKA-dependent phosphorylation. This transient period of Rho activation correlates with ongoing DAT internalization. The duration of this window can be shortened in vitro and in vivo by stimulating cell-surface Gs-coupled receptors (e.g. D1/5 dopamine receptors), which raise cAMP and terminate Rho-activation. Pretreatment of mice with agents that selectively increase cAMP in dopaminergic neurons reduces AMPH-evoked behaviors without altering those induced by cocaine. These results indicate that AMPH can activate Rho, Rac, and PKA signaling independent of its action on vesicular or plasma membrane transporters. The activation of cytoplasmic signaling cascades by amphetamines contributes to the behavioral effects of acute amphetamine exposure and may explain some of the unique neurobiological consequences associated with amphetamine use and abuse.

A Retrospective on PET and the Amphetamine Paradigm: Validation and Paradoxes

Anissa Abi-Dargham

Columbia University, New York, United States

BACKGROUND: The amphetamine paradigm consists in using amphetamine administration to produce an acute change in synaptic concentrations of dopamine that can be indexed in the brain by a resulting change in the binding of $D_{2/3}$ radiotracers during a SPECT or PET experiment. This paradigm has been validated as an index of dopamine release by microdialysis studies showing a linear relationship between microdialysis measurements of extracellular dopamine and displacement of $D_{2/3}$ radiotracers by different doses of amphetamine in non human primates (1). While amphetamine has a complex set of effects at the cellular level, overall it produces non exocytic release, by increasing the net efflux of the dopamine transporter, which should converge to increase synaptic and extrasynaptic levels (2), and increase competition with radiotracer binding at the $D_{2/3}$ sites. However, we have observed consistently through multiple studies a set of subjects that show an acute paradoxical inverse change in $D_{2/3}$ radiotracer binding after amphetamine administration suggesting a decrease in synaptic dopamine. These are usually accompanied by a corresponding behavioral effect consistent with a decrease in synaptic dopamine, such as improvement in psychosis in patients with schizophrenia. A similar pattern was observed in studies of cortical amphetamine induced dopamine release, and in patients with schizophrenia compared to controls. Furthermore, a prolonged effect of amphetamine on the binding of $D_{2/3}$ radiotracers has been observed in human and non human primates, and has been attributed to internalization of the receptors following an agonist challenge. This interpretation received support from studies of affinity to internalized and surface receptors (3) showing a decrease in affinity of most radiotracers to internalized D_2 receptors, and from the lack of a late displacement effect in a rodent model of impaired internalization; the arrestin3 KO mice (4).

OBJECTIVES: In this session, we will take a critical look at the imaging data, reviewing previously published as well as unpublished new studies, focusing on aspects that have not been well explained by our common understanding of the effects of amphetamine, and set the stage for an in depth analysis of the cellular effects of amphetamine and their imaging correlates. The complexity of amphetamine's effects at the cellular level and as an imaging paradigm presents a challenge but also an opportunity as a tool to explore various aspects of dopaminergic transmission.

References

1. Laruelle M, Iyer RN, al-Tikriti MS, Zea-Ponce Y, Malison R, Zoghbi SS, et al. (1997): Microdialysis and SPECT measurements of amphetamine-induced dopamine release in nonhuman primates. *Synapse.* 25:1–14.
2. Sulzer D (2011): How addictive drugs disrupt presynaptic dopamine neurotransmission. *Neuron.* 69:628–649.
3. Guo N, Hwang DR, Lo ES, Huang YY, Laruelle M, Abi-Dargham A (2003): Dopamine depletion and in vivo binding of PET D1 receptor radioligands: implications for imaging studies in schizophrenia. *Neuropsychopharmacology.* 28:1703–1711.
4. Skinbjerg M, Liow JS, Seneca N, Hong J, Lu S, Thorsell A, et al. (2010): D2 dopamine receptor internalization prolongs the decrease of radioligand binding after amphetamine: a PET study in a receptor internalization-deficient mouse model. *NeuroImage.* 50:1402–1407.

The Corelease of Glutamate by Dopamine Neurons

Robert Edwards

Departments of Physiology and Neurology, UCSF School of Medicine, USA

Dopamine neurons of the ventral tegmental area have been shown to form glutamatergic connections in vitro and to express the vesicular glutamate transporter VGLUT2 in vivo. To understand the physiological role of glutamate corelease by these dopamine neurons, we knocked out VGLUT2 specifically in dopamine neurons, and observed effects on vesicle filling with dopamine. Biophysical studies show that glutamate promotes a larger, more stable pH gradient than chloride, thus making a bigger driving force for vesicular monoamine transport. Differences in ionic coupling by the glutamate and chloride carriers apparently account for these differences. In addition, the expression of channelrhodopsin specifically in dopamine neurons confers light-evoked release of glutamate in the striatum, demonstrating that glutamate also acts as an independent signal.

Psychostimulant-Induced Alterations of Monoaminergic Transporters: Implications for Neurotoxicity and Neuroprotection

Annette E. Fleckenstein and Glen R. Hanson

Department of Pharmacology and Toxicology, University of Utah, Salt Lake City, UT, USA

Repeated high-dose methamphetamine (METH) administrations cause a rapid decrease in dopamine sequestration capacity via the vesicular monoamine transporter-2 (VMAT-2). Others and we have suggested that this effect is likely due to a redistribution of VMAT-2 protein, as well as associated vesicles, within dopaminergic nerve terminals. Consequently, unsequestered cytoplasmic dopamine accumulates leading to reactive species formation and persistent dopaminergic deficits. In contrast, methylphenidate increases VMAT-2-mediated dopamine sequestration, as assessed in non-plasmalemmal membrane-associated (presumably cytoplasmic) vesicles. This occurs concurrent with an increase in VMAT-2 immunoreactivity in this subcellular fraction. Methylphenidate post-treatment likewise attenuates the persistent dopaminergic deficits caused by repeated high-dose METH treatment. Mechanisms underlying, and the therapeutic implications of, these phenomena will be discussed.

Trace Amine Associated Receptor 1 is a Catecholamine Receptor that Differentially Controls DAT and NET Internalization

Gregory M Miller, Bertha K Madras and Zhihua Xie

Harvard Medical School/NEPRC, Southborough, MA, United States

BACKGROUND: Past research has shown that Trace Amine Associated Receptor 1 (TAAR1) activation alters dopamine neuron firing rates as well as modulates dopamine transporter (DAT) and norepinephrine transporter (NET) uptake and efflux functions.

OBJECTIVES: To define the role of TAAR1 signaling in DAT and NET internalization.

METHODS AND RESULTS: We used transfected (TAAR1, DAT, DAT + TAAR1, NET, NET + TAAR1) HEK293 cells, and striatal (ST) and thalamic (TH) synaptosomes from wild type (WT, 75% C57BL/6J and 25% 129S1/Sv) and TAAR1 knockout (KO; derived by Wolinsky et al., 2007) mice. Major findings: 1. TAAR1 signals through both the cAMP/PKA and PKC/Ca++ pathways in vitro: Methamphetamine (METH, 1 μM), a potent TAAR1 agonist, selectively activated both pathways, similarly in human, rhesus monkey and mouse TAAR1 cells in Cignal45 assays (Qiagen), but not in untransfected HEK293 cells. 2. TAAR1 activation by METH increased phosphorylation of both PKA and PKC in vitro and in WT/ST and WT/TH ex vivo, but not in DAT cells or KO/ST or KO/TH. 3. PKC-driven DAT internalization is TAAR1-dependent: METH induced DAT internalization in TAAR1 + DAT cells and WT/ST ex vivo, but not in DAT cells or KO/ST. These findings translated in vivo: METH (1 or 10 mg/kg) i.p. induced time-dependent DAT internalization in the excised WT/ST, but not in KO/ST. Ex vivo, DAT internalization in response to METH or PMA (1 μM) was blocked by Ro32-0432 (10 μM) but was unaffected by 8-Br-cAMP (100 μM) or H89 (10 μM). 4. Unlike DAT, PMA but not METH induced NET internalization, and 8-Br-cAMP (100 μM) reversed this effect. 5. Notably, METH induced robust NET internalization when the PKA pathway was concurrently inhibited by H89, in a TAAR1-dependent manner.

CONCLUSIONS: We present a novel mechanism by which METH and other TAAR1 drugs can affect DAT and NET internalization, with therapeutic implications for neuropsychiatric/addictive disorders.

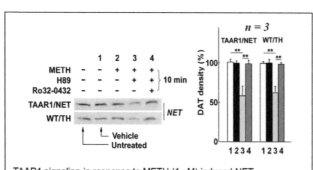

TAAR1 signaling in response to METH (1 μM) induced NET internalization when PKA was inhibited by H89 (10 μM) *in vitro* and *ex vivo*, and this effect was blocked by PKC inhibition (10 μM Ro32-0432). **p < 0.01 by one-way ANOVA.

VGF-Derived Peptide TLQP-21: Structure, Function, and Biomedical Implications

Cheryl Cero[1], Raffaello Verardi[2], Maria Razzoli[1], Alessio Lodola[3], Roberta Possenti[4], Gianluigi Veglia[2] and Alessandro Bartolomucci[1,*]

[1]Dept. Integrative Biol. Physiol; [2]Dept. Biochem. Mol. Biol., University of Minnesota, Minneapolis, MN 55455, USA; [3]Pharmaceutical Dept., University of Parma, Parma, 43100 Italy; [4]Dept. Medicine of System, University of Rome 'Tor Vergata', Rome, Italy; *Corresponding author: abartolo@umn.edu

VGF is a member of the granin family of peptides [1]. The VGF gene encodes a 615 amino acid protein (617 in rodents) stored in dense core granules and processed to several bioactive peptides, released through the regulated pathway. At least 7 peptides possess non-redundant biological functions in rodents, and VGF fragments are becoming useful biomarkers for human disease [1]. We identified and characterized the C-terminal internal fragment designated TLQP-21 [2], a multifunctional peptide active in regulating energy balance, gastric function, nociception, reproduction and stress [1]. Intracerebroventricular infusion of TLQP-21 increases energy expenditure and prevents obesity without activating central hypothalamic neuropeptides or feeding [2]. In periphery, TLQP-21 [3] is co-stored with tyrosine hydroxylase in sympathetic nerves innervating the adipose organ; increases lipolysis downstream of β-adrenergic receptor signaling; increases sympathetic innervation to adipose fat pads; and decreases adipocyte diameters without exerting negative cardiovascular effects or activating energy expenditure [3]. We used solution and solid-state NMR spectroscopy to investigate the conformational structure of TLQP-21 in free and receptor-bound state. In solution [3], TLQP-21 does not possess a defined secondary structure and is present in equilibrium among several conformations. Solid-state NMR analysis of the peptide in the presence of 3T3L1 cells (expressing its putative receptor [3]) demonstrate a conformational change and acquisition of a α-helix structure as demonstrated by significant Cα and Hα chemical shifts. The elucidation of the structure-function relationships of TLQP-21 and its interaction with the receptor represent crucial steps towards how TLQP-21 signals, and designing small molecules for the treatment of obesity.

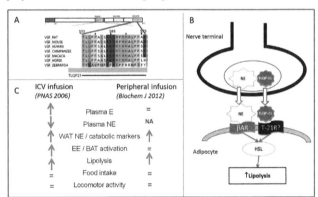

Acknowledgements

Supported by Minnesota Partnership for Biotech. Medical Genom., Decade of Discovery in Diabetes Grant (AB).

References

1. Bartolomucci et al., *Endocr Rev.* 2011;32:755–97
2. Bartolomucci et al., *Proc Natl Acad Sci USA.* 2006;103:14584–9
3. Possenti et al., *Biochem J.* 2012;441:511–22.

Chromogranins the Key Proteins for the Storage and Regulation of Exocytosis in Chromaffin Cells

Natalia Domínguez[1], José G. Hernández-Jiménez[2], María-Rosa Arnau[1], José-David Machado[1] and Ricardo Borges[1,2]

[1]*Pharmacology Unit, Medical School;* [2]*IUBO-AG, La Laguna University, Tenerife, Spain*

BACKGROUND: Chromogranins are acidic secretory proteins, which are primarily found in the dense-core secretory vesicles (LDCV) of neuroendocrine cells and neurons. The most abundant members are chromogranin A (CgA), chromogranin B (CgB) and secretogranin II (SgII). Their functions are still debated but evidences point in four main directions: LDCV biogenesis, prohormone sorting and packaging of neuropeptides in granules within the trans-Golgi network, the biological activities of their derivative peptides, and the storage of transmitters such as catecholamines (CA) and ATP.

OBJECTIVES: We have focused our research in exploring the contribution of Cgs in the amine cargo and exocytosis.

METHODS AND RESULTS: Using amperometry, patch-amperometry and intracellular electrochemistry in chromaffin cells from Cgs-KO mice we have demonstrated that the absence of CgA and CgB caused important changes in CA accumulation and in the kinetics of exocytosis using a CgA- and CgB-KO mice. We have obtained a double CgA/B-KO mouse by crossing those strains, which resulted viable and fertile. The CA content in chromaffin LDCV is halved and the secretory response largely reduced. Incubation with L-DOPA increased the vesicle CA content of WT- but not in Cgs-KO cells. This was not due to changes in amine transport nor in the synthesis or degradation of cytosolic amines. Electron microscopy revealed the presence of giant and highly altered secretory vesicles with little electrodense inner matrix. Proteomic analysis confirmed the absence of CgA&B, little changes in SgII in the LDCV-enriched fraction, but the overexpression of fibrinogen and other proteins.

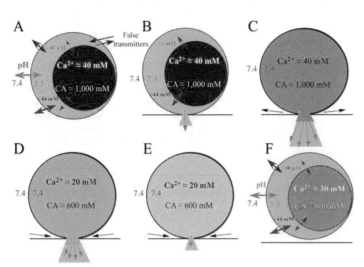

Supported by MICINN (BFU2010-15822) and CONSOLIDER (CSD2008-00005) and ACIISI/FEDER PI2007/017&C2008/01000239.

INTERPRETATION AND CONCLUSIONS: Cgs are crucial for the aggregation of vesicular content allowing the high concentration of CA and other soluble species. This association is largely responsible for the kinetic of exocytosis found at single event level.

Chromogranin A in the Storage and Exocytosis of Catecholamines

**Natalia Dominguez[1], Judith Estevez-Herrera[1], Josue Campos[1],
Ricardo Borges[1,2] and Jose D. Machado[1]**

[1]Pharmacology Unit, Medical School; [2]IUBO-AG, La Laguna University, Tenerife, Spain

BACKGROUND: The chromogranins (Cgs) are highly acidic secretory proteins that are primarily found in the large dense-core secretory vesicles (LDCV) of neuroendocrine cells and neurons, being stored with neurotransmitters, hormones and peptides. The most abundant members in chromaffin cells are chromogranin A (CgA), chromogranin B (CgB) and secretogranin II. The evidences point towards to four main functions: LDCV biogenesis; prohormone sorting and packaging of neuropeptides; storage of transmitters such as catecholamines (CA); and the biological activities of their derivative peptides. In chromaffin cells from Cgs-KO mice, we have reported that the absence of CgA and CgB causes important decline in the CA accumulation and in the kinetics of exocytosis (Montesinos, 2008, Díaz-Vera, 2010, Díaz-Vera, 2012). Therefore Cgs are a highly efficient system directly implicated in monoamine cargo and in the kinetics of exocytosis.

OBJECTIVES: We wanted to study: i) expression of newly formed CgA; ii) role of CgA in the formation and sorting of LDCV and iii) participation of CgA in the storage and exocytosis of CA.

METHODS AND RESULTS: CgA-EGFP expression in HEK 293 and PC12 cells. CgA-EGFP was accumulated in granule-like subcellular structures. This expression was confirmed by Western Blot (WB) and analyzed using TIRFM. CgA-EGFP is released by exocytosis. Using TIRFM, we quantified the exocytotic response. HEK293-CgAEGFP cells accumulate L-DOPA. Using HPLC we measured an increase of L-DOPA accumulated. Quantal-release of CA: Amperometric recordings could detect quantal release from HEK293-CgA-EGFP.

CONCLUSIONS: CgA-EGFP has the capability to generate granule-like vesicles in both cell lines, which are able to undergo exocytosis thus releasing CgA-EGFP. These granule-like structures are able to storage and release L-DOPA/DA by exocytosis.

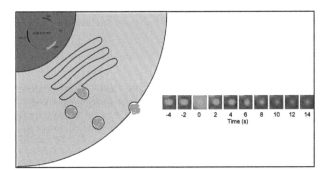

ND and JEH are recipients of FPU and FPI respectively from ME and MICINN. JDM has a Ramon y Cajal contract. Supported by MICINN (BFU2010-15822) and CONSOLIDER (CSD2008-00005) (RB).

References

1. Montesinos MS, Machado JD, Camacho M, Diaz J, Morales YG, Alvarez de la Rosa D, Carmona E, Castaneyra A, Viveros OH, O'Connor DT *et al*: **The crucial role of chromogranins in storage and exocytosis revealed using chromaffin cells from chromogranin A null mouse.** *J Neurosci* 2008, **28**(13):3350–3358.

2. Díaz-Vera J, G. Morales Y, Hernández-Fernaud JR, Camacho M, Montesinos MS, Calegari F, Huttner WB, Borges Ra, Machado JD: **Chromogranin B gene ablation reduces the catecholamine cargo and decelerates exocytosis in chromaffin secretory vesicles.** *Journal of Neuroscience* 2010.

3. Diaz-Vera J, Camacho M, Machado J, Dominguez N, Montesinos M, Hernandez-Fernaud J, Lujan R, Borges R: **Chromogranins A and B are key proteins in amine accumulation but the catecholamine secretory pathway is conserved without them.** *FASEB J* 2012, **26**:430–438.

The Involvement of Vesicular ATP in the Storage and Exocytosis of Catecholamines of Bovine Chromaffin Cells

Judith Estévez-Herrera[1], Marta R. Pardo, Natalia Domínguez[1], Carmen Jiménez-Espinoza[2], Ricardo Borges[1,2] and José D. Machado[1]

[1]*Pharmacology Unit, Medical School;* [2]*IUBO-AG, La Laguna University, Tenerife, Spain*

BACKGROUND: Chromaffin granules are organelles similar to the large dense core vesicles (LDCV) present in many secretory cell types including neurons. They accumulate solutes at high concentrations (catecholamines, 0.5–1 M; ATP, 120–300 mM; or Ca^{2+}, 40 mM) (Winkler & Westhead, 1980) and can achieve a theoretical osmolarity around 1.5-2 mOsm, a tonicity much greater than cytosol.

The co-storage of ATP and catecholamines seems to reduce the effective intragranular osmolarity to that of the cytoplasm thus preventing LDCV lysis (Kopell & Westhead, 1982). This contribution leads to the idea that ATP forms a complex with vesicular catecholamines stabilizing their high intravesicular concentration. Recently, a nucleotide transporter present in secretory vesicles known as VNUT has been identified (Moriyama, 2008).

OBJECTIVES: We wanted to know the role of VNUT in the vesicular accumulation of ATP and the consequences of its inhibition in the catecholamine storage and exocytosis.

METHODS AND RESULTS: VNUT-EGFP expression is reduced by siRNA in HEK 293T cells, as was shown using flow cytometry and Western blot.

VNUT is responsible for vesicular storage and subsequent exocytosis of ATP in chromaffin cells. The silencing of VNUT caused a decrease in SLC17A9 mRNA expression that was quantified by qRT-PCR. The nicotinic-triggered secretion of ATP resulted largely diminished in cells, which VNUT was silenced.

Catecholamine secretion is reduced when VNUT is silenced. HPLC analysis shows a reduction in triggered secretion when compared to control chromaffin cells.

Amperometric recordings were used to quantify the kinetics of exocytosis and quantum size. Preliminary data show a slightly reduction in the quantum size and a slowdown of exocytosis.

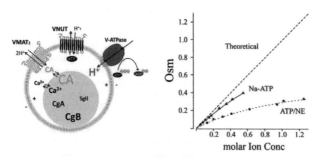

Supported by MICINN (BFU2010-15822) and CONSOLIDER (CSD2008-00005) and ACIISI/FEDER PI2007/017&C2008/01000239.

INTERPRETATION AND CONCLUSIONS: Our results supported the participation of VNUT in vesicular storage of ATP in the secretory vesicles of chromaffin cells. The observed reduction of VNUT also affected the releasable ATP and influenced the exocytosis of catecholamines.

References

Winkler H, Westhead E: **The molecular organization of adrenal chromaffin granules.** *Neuroscience* 1980, **5** (11):1803–1823.

Kopell WN, Westhead EW: **Osmotic pressures of solutions of ATP and catecholamines relating to storage in chromaffin granules.** *J Biol Chem* 1982, **257** (10):5707–5710.

The Role of Ca^{2+} Release from the Endoplasmic Reticulum in the Catecholamine Hypersecretion of Adrenal Chromaffin Cells from Spontaneously Hypertensive Rats

López-Bistrain P., Segura-Chama P., Jiménez-Pérez N., Rivera-Cerecedo C.V. and Hernández-Cruz A.

Instituto de Fisiología Celular, Universidad Nacional Autónoma de México, México City, D.F. CP 04310. México.

BACKGROUND: Enhanced catecholamine (CA) release from adrenal chromaffin cells (CCs) of spontaneously hypertensive rats (SHR) could result from increased Ca^{2+} entry through voltage-gated Ca^{2+} channels and/or Ca^{2+} release from intracellular stores. Since Ca^{2+} currents in SHR and normotensive Wistar Kyoto rat (WKY) CCs are quite similar, a greater participation of intracellular Ca^{2+} stores in SHR CCs is likely. Interestingly, spontaneous fluctuations of intracellular Ca^{2+} are larger and more frequent in SHR compared to WKY CCs (unpublished observations).

OBJECTIVE: Characterize in cultured CCs from SHR and WKY rats, both basal and stimulated CA secretion using carbon fiber amperometry and evaluate the contribution of Ca^{2+} release from intracellular stores.

METHODS AND RESULTS: Blood pressure (BP) was measured via tail-cuff plethysmography in male SHR and WKY rats 7−12 weeks old. SHR had BP values significantly greater than WKY rats (systolic (mm Hg: 147.1 *vs.* 111.2 and diastolic: 108.7 *vs.* 81.4: respectively). Cultured CCs were used 24−48 h after plating. For amperometric recordings carbon fiber electrodes kept at +700 mV were used. CCs were perfused with Krebs-HEPES solution to evaluate spontaneous secretory events and then CA secretion was stimulated by local application of high-K$^+$ (60 mM for 5 s) or caffeine (10 mM for 10 s). 73% of WKY CCs and 100% of SHR CCs showed spontaneous CA release events (figs. 1,2). Amperometric spikes elicited by high-K$^+$ and caffeine are larger in amplitude and mean charge in SHR compared to WKY CCs (figs. 1,2 and table). Total integral charge reached over time is also significantly greater in SHR CCs (fig. 3) in spite of their briefer amperometric spikes (see table).

CONCLUSIONS: The comparison of secretory events in SHR and WKY CCs strongly suggest a greater contribution of Ca^{2+}-induced Ca^{2+} release from intracellular stores in SHR compared to WKY CCs.

- **Amperometric spike recordings in WKY and SHR CCs**

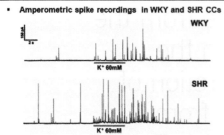

Fig.1. CCs were perfused with Krebs-Hepes solution for 120 s to record spontaneous secretory events. The recording continued for 120 s and then a 5 s pulse of high-K+ was applied, which elicited a more rapid and longer-lasting burst of secretory spikes in CCs SHR during the first 10 s when compared to WKY cells.

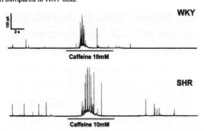

Fig.2. CCs were perfused with Krebs-Hepes solution to record spontaneous secretory events. Then 10 s pulse of 10 mM caffeine was applied which elicited a burst of spikes with a delay of approximately 2 s noticed for both CCs strains. SHR cells responded with a stronger secretory activity compared to WKY cells.

- **Cumulative integral charge of amperometric spikes**

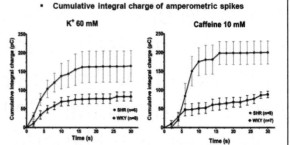

Fig.3. Comparison of the cumulative integral charge (Q) of WKY and SHR CCs following a pulse of either high-K+ or caffeine. High-K+ perfusion elicited spikes with a cumulative integral charge 2-fold higher in SHR than WKY CCs (162 ± 14.9 pC and 75.97 ± 13.1 pC, respectively; p<0.001). Caffeine application elicited spikes with a cumulative integral charge 3-fold higher in SHR CCs compared to WKY CCs (198.65 ± 30.3 pC and 62.5 ± 16 pC, respectively; p<0.001). Data: Mean ± S.E.M

Kinetic parameters evaluated for individual amperometric spikes in CCs from both rat strains:

	K+ 60 mM			Caffeine 10 mM		
	I_{max} (pA)	$t_{1/2}$ (ms)	Integral Charge (pC)	I_{max} (pA)	$t_{1/2}$ (ms)	Integral Charge (pC)
WKY	25.4±1.7	10.5±0.4	4.3±0.3	48.4±3.6	10.8±0.5	6.3±0.4
SHR	62±4.8*	8.4±0.4*	7±0.6**	94.3±6.9*	8.6±0.4*	10.3±0.7**

Data are Means ± SEM, *p<0.05; **p<0.001

Supported by Grants PICSA10-116 (ICyTDF), CONACyT 79763 and DGAPA-PAPIIT IN2279.

PACAP as a Primary Effector of the Acute Sympatho-Adrenal Stress Response

Corey Smith

Case Western Reserve University, Cleveland, OH, USA

BACKGROUND: Catecholamine-secreting adrenal medullary chromaffin cells receive excitatory synaptic input from the sympathetic splanchnic nerve. Under basal sympathetic tone experienced at rest, the splanchnic terminals release acetylcholine to fine-tune low-level adrenal excitation. Sympathetic stress signals persistent and robust adrenal excitation through splanchnic release of PACAP, a potent adrenal secretagogue.

OBJECTIVES: We set out to characterize PACAP-mediated stimulation that leads to initial catecholamine secretion from single chromaffin cells as well as tissue-level electrical remodeling to elevate overall adrenal medullary excitation.

METHODS AND RESULTS: Catecholamine release was measured in a tissue slice preparation by single-cell electrochemical amperometry and cell electrical activity by perforated patch voltage/current clamp. Exogenous PACAP (1 μM) application or burst-mode bipolar stimulation of the innervating splanchnic revealed Ca^{2+} influx and robust catecholamine exocytosis from chromaffin cells. Pharmacologic and molecular perturbation revealed PACAP-evoked secretion to include an Epac-mediated activation of protein kinase C (PKC), leading to a phosphorylation-dependent recruitment of low voltage-activated T-type calcium channels to a conductive state as well as a membrane depolarization to approximately -50 mV. This potential is subthreshold for action potential firing but sufficient to gate T-type Ca^{2+} channels. In parallel, PKC activation elevates gap junction coupling and conductance between chromaffin cells. Such gap junction coupling is expected to result in the spread of excitation between the heterogeneously-innervated chromaffin cells to increase overall medullary excitation, and thus, increase stress-evoked catecholamine release.

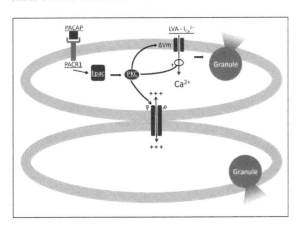

CONCLUSIONS: Splanchnic release of PACAP represents a critical step in the sympatho-adrenal stress reflex and supports catecholamine release to meet the metabolic response to acute "fight-or-flight" stress. PACAP excitation of the adrenal medulla is separate from cholinergic excitation and occurs through an activation of a second messenger pathway to decrease membrane potential and recruit low voltage-activated Ca^{2+} channels. At the tissue level, PACAP evokes elevated intercellular electrical coupling to increase overall adrenal catecholamine release.

Amphetamine Induced Dopamine Release in Macaque Monkeys: Striking Differences Between Prefrontal Cortex and Caudate Nucleus

Charles W. Bradberry[1,3], Hank P. Jedema[1], Kate Gurnsey[1] and Raj Narendran[1,2]

[1]*Depts. of Psychiatry;* [2]*Radiology;* [3]*University of Pittsburgh, and Pittsburgh VA Health Services, Pittsburgh, PA*

BACKGROUND: There is a need for methods that allow the clinical investigation of stimulated cortical dopamine (DA) release, similar to existing PET based methods employing displacement of DA receptor radioligands in the striatum using amphetamine. There is currently very little known about the relative effects of amphetamine on cortical versus striatal function in primates.

OBJECTIVES: To validate an imaging approach to study cortical DA release through amphetamine-induced displacement of radiotracers from dopaminergic receptors.

METHODS AND RESULTS: As part of validation studies for the use of the PET ligand [^{11}C]FLB 457 displacement for measuring amphetamine-induced DA release, we conducted a dose dependent comparison of the ability of amphetamine to increase extracellular DA in prefrontal cortex and striatum using microdialysis during simultaneous PET studies. Amphetamine led to a significant dose-dependent increase in extracellular cortical DA as measured with microdialysis and decrease in [^{11}C]FLB 457 binding potential (BP$_{ND}$) as measured with PET. The relationship between amphetamine-induced peak ΔECF DA and Δ[^{11}C]FLB 457 BP$_{ND}$ in the frontal cortex and striatum was linear, but the slope of the dose-peak response relationship in the caudate was approximately 3.4 times steeper than the slope in the cortex. There was also a striking difference in the temporal profile, with the striatal changes peaking at 20 min post-injection, and then falling rapidly, whereas cortical DA levels were not maximal until 30−40 min post injection, and thereafter decayed slowly.

CONCLUSIONS: The results of this study clearly demonstrate that the magnitude of extracellular dopamine release is correlated with the magnitude of the reduction in [^{11}C]FLB 457 BP$_{ND}$ in the frontal cortex, indicating suitability for use in clinical studies. They also suggest substantial differences in the timecourse and magnitude of amphetamine effects on extracellular DA in cortex and striatum, consistent with differences in autoregulatory mechanisms between their respective DA projections.

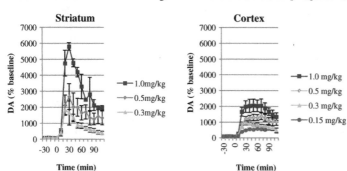

Attenuated Amphetamine-Induced Dopamine Release in Subjects at Ultra-High Risk for Substance Dependence

Kevin F. Casey[1], Chawki Benkelfat[1,2], Mariya V. Cherkasova[3], Alain Dagher[2] and Marco Leyton[1,2,3]

[1]*Department Psychiatry, McGill University, Montreal, QC, Canada;* [2]*Department of Neurology & Neurosurgery, McGill University, Montreal, QC, Canada;* [3]*Department of Psychology, McGill University, Montreal, QC, Canada*

BACKGROUND: Drugs of abuse increase dopamine transmission. Indirect evidence from animal models and studies in people with current addictions suggest that this response might be disturbed in those at risk for substance use disorders.

OBJECTIVES: Measure drug-induced striatal dopamine responses in young adults with a multigenerational family history (FH-Pos) of addictions.

METHODS AND RESULTS: Three groups were tested. 1) FH-Pos: 16 non-dependent subjects (age: 21.3 ± 2.4 y. o.) with a multigenerational FH of substance abuse and a personal history of cocaine or amphetamine use (49.6 ± 59.3 stimulant uses). 2) FH-Neg: 15 subjects (22.1 ± 1.8 y.o.) with no 1st or 2nd degree relatives with substance use problems, matched to FH-Pos on their personal history of drug use (43.5 ± 40.7 stimulant uses). 3) Control (CTRL): 17 stimulant drug naïve healthy controls without substance abusing relatives (20.1 ± 1.8 y.o.). All subjects had two PET [^{11}C]raclopride scans, one with d-amphetamine (0.3 mg/kg p.o.), one with placebo, given double-blind.

FH-Pos subjects had a smaller [^{11}C]raclopride response to amphetamine than either FH-Neg or CTRL groups, and these two control groups did not differ from one another. The differential response was most pronounced in the right ventral striatum where amphetamine decreased [^{11}C]raclopride binding values by $15.7 \pm 15.0\%$ in CTRL, $14.3 \pm 8.4\%$ in FH-Neg, and only $3.4 \pm 12.6\%$ in FH-Pos ($p = 0.006$ *vs.* CTRL, $p = 0.021$ *vs.* FH-Neg). Amphetamine-induced pleasurable effects were greater in the subjects at risk for addiction ($p = 0.021$ *vs.* CTRL, $p = 0.029$ *vs.* FH-Neg).

CONCLUSIONS: The present study suggests that subjects at familial risk for addiction have diminished dopaminergic responses to d-amphetamine. The effect remained after controlling for past drug use, was accompanied by elevated pleasurable effects, and resembled that reported in currently dependent alcoholics and cocaine addicts. Since the present study's subjects were non-dependent, low dopamine responsivity appears to express itself well before extensive drug exposure, potentially reflecting a trait marker for addiction.

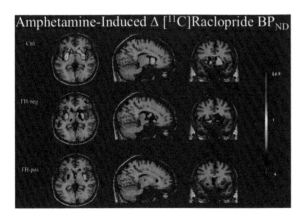

Amphetamine-Induced Δ [^{11}C]Raclopride BP$_{ND}$

The Membrane-Raft Protein Flotillin-1 is Essential in Dopamine Neurons for Amphetamine-Induced Behavior in Drosophila

Andrea B. Pizzo[1], Caline S. Karam[1], Yuchao Zhang[1], Ai Yamamoto[2], Brian D. McCabe[3] and Jonathan A. Javitch[1,4]

[1]*Departments of Psychiatry;* [2]*Neurology and;* [3]*Neuroscience, Columbia University College of Physicians and Surgeons, New York, NY, USA;* [4]*Division of Molecular Therapeutics, New York State Psychiatric Institute, New York, NY, USA.*

BACKGROUND: The dopamine transporter (DAT) is the primary molecular target for the rewarding properties of psychostimulants. Amphetamine (AMPH) increases extracellular dopamine (DA) by promoting its non-exocytotic release via DAT-mediated efflux. Studies in heterologous cells have shown that phosphorylation of DAT and its localization in membrane rafts are required for AMPH-induced DA efflux but not for DA uptake. However, the mechanisms that coordinate efflux and the ensuing behavioral effects remain poorly defined.

OBJECTIVES: We developed a behavioral assay for the effects of psychostimulants *in vivo* using *Drosophila melanogaster.* Key mechanisms of neurotransmission are conserved between flies and humans and given its accessibility to genetic analyses, the fly provides a powerful model to address these phenomena broadly and efficiently.

METHODS AND RESULTS: Larvae were treated with either vehicle or a drug solution and their crawling speed was measured using automated methods. Larvae fed acute doses of AMPH respond with an increase in their crawling velocity. Using a variety of genetic approaches, we demonstrate that this behavioral response is dependent on DAT and its phosphorylation. We also show that methylphenidate (MPH), which competitively inhibits DA uptake but does not induce DA efflux, also leads to DAT-dependent hyperlocomotion, but this response is independent of DAT phosphorylation. We also demonstrate that the raft protein Flotillin1 (Flot1) is required for AMPH-induced but not MPH-induced hyperlocomotion. These results are the first evidence of a role for a raft protein in an AMPH-mediated behavior and are consistent with a model in which Flot1 promotes AMPH-induced DA efflux by localizing DAT to membrane microdomains where it can be phosphorylated by co-localized kinases. Thus, using our assay we are able to translate molecular and cellular findings to a behavioral level and to differentiate *in vivo* the distinct mechanisms of two psychostimulants.

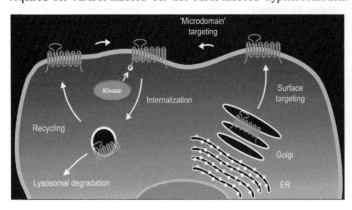

Plant Flavonoids Affecting Adrenal Medullary Catecholamine Biosynthesis and Release

Nobuyuki Yanagihara[1], Yumiko Toyohira[1], Susumu Ueno[1], Masato Tsutsui[2] and Han Zhang[3]

[1]Department of Pharmacology, University of Occupational & Environmental Health, School of Medicine, Kitakyushu, 807-8555, Japan; [2]Department of Pharmacology, Faculty of Medicine, University of The Ryukyus, Okinawa, 903-0215, Japan; [3]Research Center of Traditional Chinese Medicine, Tianjin University of Traditional Chinese Medicine, Tianjin, China

BACKGROUND: Recent research attention has been paid to high dietary intake of plant flavonoids because of their potentially beneficial effects associated with a reduction in the risks of age and life-style related diseases such as cardiovascular diseases, diabetes and cancers. Resently, we reported that several flavonoids such as daidzein, a soy isoflavone, and resveratrol, a grape polyphenol, stimulated catecholamine (CA) synthesis and tyrosine hydroxylase activity mediated through plasma membrane estrogen receptors.

OBJECTIVES: We report here the view on the insights into pharmacological potential of plant flavonoids such as nobiletin and daidzein on catecholamine signaling in cultured bovine adrenal medullary cells used as a model of the sympathetic nervous system.

METHODS AND RESULTS: Treatment of cultured bovine adrenal medullary cells with nobiletin and daidzein stimulated ^{14}C-catecholamine synthesis from [^{14}C]tyrosine and tyrosine hydroxylase activity. The stimulatory effects of nobiletin and daidzein on catecholamine synthesis were abolished by derivation of extracellular Ca^{2+} or partially inhibited by H-89, an inhibitor of cyclic AMP-dependent protein kinase. Nobiletin (1.0–100 μM) also concentration-dependently stimulated catecholamine secretion and $^{45}Ca^{2+}$ influx. The stimulatory effect of nobiletin on catecholamine secretion was blocked by deprivation of extracellular Ca^{2+} or by specific inhibitors (L-type and N-type) of voltage-dependent Ca^{2+} channels. On the other hand, acetylcholine, a physiological secretagogue, increased the secretion of catecholamines, which was inhibited by nobiletin and daidzein (1–100 μM). Nobiletin also suppressed catecholamine secretion and $^{22}Na^+$ and $^{45}Ca^{2+}$ influx induced by acetylcholine.

CONCLUSIONS: The present findings suggest that flavonoids, nobiletin and daidzein, stimulate catecholamine synthesis and secretion via activation of a Ca^{2+}- and/or PKA-dependent pathway, whereas they suppress catecholamine secretion through inhibition of nACh receptor-ion channels in bovine adrenal medullary cells. Supported by Grant-in-Aids (23617035, 23590159, and 23617036) for Scientific Research (C) from the Japan Society for the Promotion of Science.

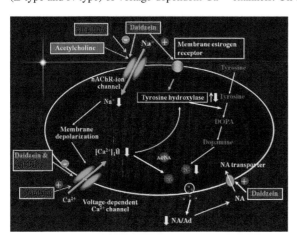

A Protein Complex for Dopamine Synthesis and Vesicular Storage: Target for Cellular Oxidative Damage

Tracy Baust[1] and Gonzalo E. Torres

[1]*University of Pittsburgh, Pittsburgh, Pennsylvania, United States*

BACKGROUND: The dopamine (DA) system has been implicated in the pathophysiology of many neurological and psychiatric conditions, including Parkinson's disease. Central to the involvement of the DA system in brain disorders is the idea that DA neurons are especially vulnerable to insult due to their neurotransmitter's ready ability to auto-oxidize and form dangerous quinones and free radicals. Since the formation of free radicals occurs in the cytosol, there is an important need for the neuron to appropriately sequester DA. A predominant mechanism by which this occurs is through the action of the vesicular monoamine transporter 2 (VMAT$_2$), which packages DA into synaptic vesicles. DA homeostasis relies on a finely orchestrated interplay between the mechanisms responsible for DA synthesis, vesicular refilling, release, uptake, as well as degradation. Currently, many of the molecular details underlying DA synthesis and storage remain unclear.

OBJECTIVES: To examine the function and regulation of the protein complex involving DA synthesis and vesicular storage under physiological conditions as well as during the dopaminergic neurodegeneration.

METHODS AND RESULTS: We discovered an interaction between VMAT2, TH, and AADC that physically and functionally couple DA synthesis and storage at the synaptic vesicle membrane. The physiological implications of VMAT2/TH/AADC interaction on DA homeostasis may be quite important by serving to limit increases in DA levels to the local area surrounding the synaptic vesicle membrane, thus minimizing the potential oxidation of DA and the resulting toxicity (Figure). We also identified a direct interaction between the molecular chaperone heat shock cognate 70 (Hsc70) and VMAT2. Hsc70 also bids TH and promotes the targeting of the enzyme to synaptic vesicles.

CONCLUSIONS: Our results suggest a role for Hsc70 in DA synthesis as well as its coupling to the DA storage system further implicating Hsc70 as a critical player in DA homeostasis regulation.

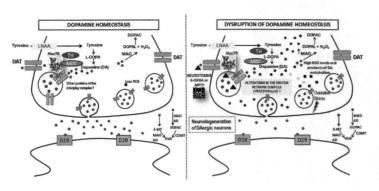

Overexpression of Parkin in The Nigrostriatal System Might Protect Dopaminergic Terminals From Methamphetamine Neurotoxicity Via Regulation of Dopamine Transporter

Bin Liu[1], Bryan Killinger[1] and Anna Moszczynska[1]

[1]Wayne State University, Detroit, MI, United States

BACKGROUND: METH is a widely abused psychostimulant, toxic at high doses to striatal dopaminergic (DAergic) terminals in experimental animals and humans. Despite years of active research in the area of METH neurotoxicity there are no specific medications that counteract the damaging effects of METH on the brain and new drug targets are needed. Parkin is capable of neuroprotection of DA neurons, suggesting the importance of parkin in DA neuron function and maintenance, which may include regulation of DA transporter (DAT). Parkin mutation or deficit in DA neurons leads to their neurodegeneration. We showed that METH caused a deficit in parkin, suggesting that this deficit mediates neurotoxicity of the drug.

OBJECTIVES: The objectives of the present study were: (1) to determine whether overexpression of parkin protects striatal DAergic terminals from high-dose binge METH *in vivo* and (2) to examine whether parkin-mediated protective mechanisms included regulation of DAT.

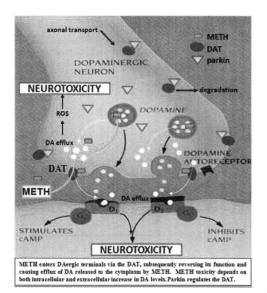

METH enters DAergic terminals via the DAT, subsequently reversing its function and causing efflux of DA released to the cytoplasm by METH. METH toxicity depends on both intracellular and extracellular increase in DA levels. Parkin regulates the DAT.

METHODS AND RESULTS: Overexpression of parkin in Sprague-Dawley rats was achieved by stereotaxic microinjection of adeno-associated viral 2/6 vector-parkin construct (AAV2/6-parkin) into the left substantia nigra. After 21 days, striatal synaptosomes were examined for the levels of DAT. In a separate group of parkin overexpressing rats, DAT levels were measured 1 h after toxic METH (7.5 mg/kg, every 2 h x 4, i.p.) or saline (1 ml/kg). Overexpression of parkin was neuroprotective. Overexpression of parkin decreased DAT in METH-naïve rats and increased DAT in METH-treated rats.

CONCLUSIONS: Overexpression of parkin might attenuate METH toxicity via decreasing DAT levels at the plasma membrane in drug-naïve rats thus letting less METH to enter the terminals in the beginning and/or by increasing DAT levels at the plasma membrane within 7 h since first METH injection thus attenuating DA efflux into the synaptic cleft.

Increased Catecholamine Toxicity in Mice with Low Expression of VMAT2

Gary W. Miller, PhD.

Emory University

BACKGROUND: Dopamine is transported into synaptic vesicles by the vesicular monoamine transporter (VMAT2; SLC18A2). Disruption of dopamine storage has been hypothesized to damage the dopamine neurons that are lost in Parkinson's disease. By disrupting vesicular storage of dopamine and other monoamines, we have created a progressive mouse model of PD that exhibits catecholamine neuron loss in the SNc and locus ceruleus and motor and nonmotor symptoms of the disease.

OBJECTIVES: To determine if reduced vesicular storage of catecholamines induces Parkinson's disease-related pathology.

METHODS AND RESULTS: Mice with a 95% reduction in VMAT2 expression and α-synuclein null were originally obtained from Piers Emson. Our laboratory restored α-synuclein through breeding with wt mice. VMAT2-deficient animals have increased formation of oxidative dopamine metabolites, increased markers of oxidative stress, and progressive damage to the dopamine system (see below). This is accompanied by decreased motor function (see below), deficits in olfactory discrimination, shorter latency to behavioral signs of sleep, delayed gastric emptying, and a progressive depressive-like phenotype. Pathologically, the VMAT2-deficient mice display progressive neurodegeneration in the SNc and the LC, coupled with α-synuclein accumulation. The mice also exhibit cardiac sympathetic dennervation.

INTERPRETATION AND CONCLUSIONS: Taken together, these studies demonstrate that reduced vesicular storage of catecholamines and the resulting disruption of the cytosolic environment may play a role in the pathogenesis of parkinsonian symptoms and neurodegeneration. In an attempt to better understand how to protect these neurons from cytoplasmic catecholamine toxicity, we have generated mice that overexpress VMAT2 (see below). These two strains of mice provide an opportunity to study the impact of reduced or elevated vesicular storage of catecholamines on a variety of neurotoxic insults and disease states.

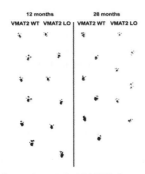

Stride length in VMAT2LO mice.

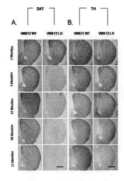

Decreased DAT and TH staining.

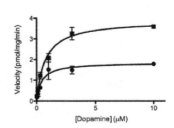

Increase dopamine uptake in BAC-VMAT2.

Metabolism

David Goldstein and Lee Eiden

Steps in catecholamine metabolism were laid out several decades ago, with monoamine oxidase (MAO) dominating intra-neuronal metabolism and catechol-O-methyltransferase (COMT) extra-neuronal metabolism. The immediate product of MAO acting on dopamine (DA) is the catecholaldehyde, 3,4-dihydroxyphenylacetaldehyde (DOPAL), which is detoxified by aldehyde dehydrogenase (ALDH). MAO and COMT figure prominently in current research on mechanisms and therapeutics of a variety of catecholamine-related diseases. Sessions of the XICS within this theme dealt with Traditional and Novel Pathways of Catecholamine Metabolism, and Catecholamine Metabolism as a Therapeutic Target in PD.

Eisenhofer et al. reviewed current understanding about catecholamine metabolism as a guide for assessment of sympathoadrenal function and dysfunction in health and disease. It was pointed out that the concept that most NE turnover occurs within sympathetic nerves due to leakage of the transmitter from vesicular stores was proposed in the 1960s by Irwin J. Kopin, one of the honorees at the XICS. At rest, over 85% of the NE produced in sympathetic nerves is metabolized intraneuronally by MAO and at least 80% of this is due to leakage of NE from vesicular stores with the remainder reflecting metabolism after reuptake. In contrast, enzymatic O-methylation represents a minor pathway of endogenous catecholamine metabolism—but with an important exception, the metabolism of catecholamines in adrenomedullary chromaffin cells—the basis for high sensitivity of plasma metanephrines in the diagnostic evaluation of symptomatic pheochromocytoma. The contribution of NE leakage to turnover provides a crucial mechanism for "gearing down" the requirement for increases in catecholamine synthesis to match increases in release, thereby providing sympathetic nerves with an extended capacity for sustainable release rates. This leakage, however, also has implications for neuropathological processes involving production of toxic deaminated aldehyde metabolites, as discussed in detail within Theme E-Neurology.

Mannisto discussed emerging concepts about the role of COMT in catecholamine metabolism. COMT is encoded by one gene, in 22q11.21 in man (chr. 16 in mouse). It has two initiation sites, two mRNAs are produced, and eventually two isoforms are formed. The COMT gene is regulated by a number of factors, of which the tissue-dependent down-regulation of the two promoter sites by estrogen is best known. In mice, COMT knockout increases evoked release and prolongs metabolic breakdown of DA in the prefrontal cortex more than in striatum or accumbens. Diatchenko et al. noted that COMT is a major contributor to pain perception. COMT inhibitors are accepted adjunctive agents in PD treatment. Bonifácio et al. reported that a new COMT inhibitor, opicapone, substantially increased and prolonged systemic levodopa levels in monkeys and rats.

Shih et al. reported on the possible roles of MAO-A and MAO-B in autism. They found that MAO-A and A/B knockout mice have elevated brain levels of monoamines and exhibit behavioral and morphological changes resembling autism-spectrum disorder. MAO A/B knockout mice had more severe behavioral alterations than did MAO-A knockout mice. They also found that MAO-A knockout in cancer cells and in the host reduced the progression and metastasis of prostate cancer. Lee et al. studied a mouse model of neurofibromatosis 1 (NF1) and found that Cdca7l, encoding the R1 transcription factor, is a top candidate for the Arlm1 locus, a male-specific modifier of astrocytoma. R1 in turn plays a role in normal physiology by inhibiting MAO. Differences in the catecholamine axis between males and females might account for the sex-specificity of modifiers of astrocytoma and pheochromocytoma.

In line with concepts presented in the description of Theme E-Neurology, Strong and Wey considered whether impaired aldehyde detoxification plays a pathogenetic role in PD. Mice null for ALDH1A1 and -2 exhibited age-dependent deficits in motor performance, and intraperitoneal L-DOPA alleviated these deficits. These abnormalities were accompanied by loss of neurons immunoreactive for TH the substantia nigra, reduced DA in the striatum, and increases in biogenic aldehydes (4-hydroxynonenal (4-HNE) and 3,4-dihydroxyphenylacetaldehyde (DOPAL), leading to the conclusions that impaired detoxification of biogenic aldehydes may be important in the pathophysiology of PD and that ALDH1A1 $\times$ 2 knockout mice may be a useful animal model of PD.

Renalase: A New Player in Catecholamine Metabolism

Gary Desir

Yale University School of Medicine, USA

Renalase was identified in the course of testing the hypothesis that the kidney secretes proteins that impact cardiovascular health, but are yet to be characterized, and could represent novel therapeutic targets. We will provide a general overview of renalase biology including its genetic association with common diseases, secretion into blood, mechanism of action, interaction with the renal dopamine and epinephrine system, and potential therapeutic utility. Association of renalase single nucleotide polymorphismshave been reported with essential hypertension, stroke, heart failure, and type 1 diabetes. Renalase secretion into blood is regulated by plasma catecholamines, renal blood flow and renal nerves, and appears to take place via a non-conventional pathway. Renal denervation markedly increases kidney renalase expression and secretion into plasma, which may partly account for the hypotensive effect of denervation. Renalase's crystal structure was solved, and confirms it is a flavoprotein capable of binding NADH and NADPH. Its mechanism of action is distinct from that of MAO-A and MAO-B, since it oxidizes catecholamines (Epinephrine $\gg$ L-DOPA $>$ Dopamine $=$ Norepinephrine) to aminochrome, and the reaction rate increases $\sim$6 fold in presence of NADH. A renalase knockout (KO) mouse model has provided valuable insights into renalase physiology. The salient phenotypic characteristics of the renalase KO include normal renal function, a 25% reduction in body weight, tachycardia, moderate hypertension, a $\sim$3 fold increase in plasma and urinary catecholamines, hypophosphatemia, and marked sensitivity to both cardiac and renal ischemia. Renalase KO mice tolerate renal and cardiac ischemia poorly, and develop ischemic necrosis that is threefold more severe than that seen in WT littermates. The administration of recombinant renalase rescues the renal and cardiac phenotype, suggesting that circulating renalase has a key role in limiting the accumulation of catecholamines in interstitial fluid during ischemia. Renalase also protects against ischemic acute kidney injury in wild type mice by reducing renal tubular necrosis, apoptosis and inflammation. The chronic administration of renalase decreases ambulatory blood pressure, and prevents the development of cardiac hypertrophy in rats. Significant progress has been made in understanding the physiology and pathophysiology of renalase. The available data indicate it plays an important role in catecholamines homeostasis, and the regulation of blood pressure and cardiac function. Most importantly, preclinical studies suggest recombinant renalase may provide a novel therapeutic option for the treatment of hypertension in kidney disease, and for the prevention and treatment of ischemic cardiac and kidney injury.

Abnormal Monoamine Oxidase A VT in the Prefrontal and Anterior Cingulate Cortex During Major Depressive Episodes and High Risk States for Major Depressive Episodes

Jeffrey Meyer

University of Toronto, Canada

Major depressive disorder is a common illness with a prevalence of 2-5% and is now the third leading cause of death and disability worldwide, according to the World Health Organization. Hence pathologies of this illness related to onset are particularly important because they offer the potential for developing novel methods of prevention. Monoamine oxidase A (MAO-A) is an enzyme found on the outer mitochondrial membrane of neurons, astrocytes and glia that metabolizes monoamines, facilitates apoptosis and oxidation. In brain tissue, levels of MAO-A are highly correlated with MAO-A activity (Nelson et al. J Neurochem1979; Saura et al. J Neurosci 1992) and MAO-A VT, an index of MAO-A density, may be quantified with positron emission tomography. Work from my laboratory and others demonstrate that elevated MAO-A level in prefrontal and anterior cingulate cortex during major depressive episodes is a replicated finding (Meyer et al. Arch Gen Psych 2006, 2009, Johnson et al. Neuropsychopharmacology 2011). We also find that several environmental influences associated with greater risk for onset of major depressive episodes are associated with elevated MAO-A VT, in the prefrontal and anterior cingulate cortex. [11C] harmine positron emission tomography was applied to measure regional MAO-A VT, an index of MAO-A levels, during major depressive episodes (n = 15), during recovery from major depressive episodes (n = 18), acute alcohol withdrawal in alcohol dependent subjects (n = 16), acute cigarette withdrawal (n = 22), during postpartum blues (n = 15), and in health (n = 28). Both during major depressive episodes and high risk states for major depressive episodes, MAO-A VTwas consistently elevated, particularly in the prefrontal and anterior cingulate cortex as compared to healthy controls (magnitudes of difference between the healthy group and sad mood states ranged from 25% to 45%, all comparisons highly significant). In states in which prefrontal and anterior cingulate cortex MAO-A VTis elevated, sad mood and/or greater risk for major depressive episodes occurs. This finding is consistent across alcohol withdrawal, cigarette withdrawal, postpartum blues and major depressive disorder even though the mechanisms implicated in elevating MAO-A VTlikely differ. A key implication of these findings is that preventing the rise in MAO-A level is a potential strategy for preventing onset of illness, and maintaining recovery.

CNS Catecholamine Metabolism – Emerging Concepts

Pekka T. Männistö

Division of Pharmacology and Toxicology, Univ. Helsinki, Finland

BACKGROUND: Catechol-O-methyltransferase (COMT) is an intracellular enzyme conjugating catechol structures, like catecholamines, L-dopa and catecholestrogens. In most parts of the brain, monoamine oxidase (MAO), another intracellular enzyme, is a major player in metabolism of catecholamines, and together with an efficient reuptake (utpake$_1$) of the released catecholamines, finishes the transmission. However, several types of evidences indicate that COMT contributes to dopamine metabolism in the prefrontal cortex (PFC) much more than in the striatum. In biochemical assays, it has been shown by Karoum et al. that 3-methoxytyramine, the DA metabolite produced by COMT in the CNS, comprises even 60% of the total basal DA turnover in the rat PFC, whereas in the striatum and nucleus accumbens this figure is $<15\%$. However, under baseline conditions, Comt gene disruption does not affect the dopamine levels in striatal, cortical or hypothalamic tissues.

METHODS AND RESULTS: In *in vivo* voltammetry studies on COMT deficient mice the different role of COMT in the PFC and striatum was further conformed. A real-time analysis of evoked dopamine overflow showed that removal of dopamine was twofold slower in the PFC of mice lacking COMT than in wild-type mice, indicating that half of the dopamine decline in this brain region results from COMT-mediated enzymatic degradation. Lack of COMT did not influence dopamine overflow/decline in the dorsal striatum. COMT-deficient mice demonstrated a small (20–25%) but consistent increase in evoked dopamine release in the PFC, but not in the dorsal striatum. A further quantifying of the alternative pathways (dopamine transporter, DAT, noradrenaline transporter, NET and MAO) was done using both no-net-flux and conventional microdialysis and blockers of DAT, NET and MAO in conscious mice, with or without Comt gene disruption. In quantitative microdialysis, but not in the conventional one - dopamine levels were increased up to 60% only in the PFC of COMT-knockout mice. In conventional microdialysis studies, it was shown that selective NET and MAO inhibition increased dopamine levels in the PFC of wild-type mice by two- to fourfold, an effect that was still doubled in COMT-KO mice. Inhibition of DAT had no effect on dopamine levels in either genotype. Therefore, it was concluded that in the mouse, PFC COMT contributes about one half of the total dopamine clearance. NET and MAO eliminate about one half of released dopamine and the remaining half is cleared through sequential uptake2 and COMT. The role of DAT in prefrontal dopamine clearance is minor if any. These results indicate a particular role for COMT in prefrontal cortical dopamine metabolism compared to the striatum and NAcc.

CONCLUSIONS: Together these findings demonstrate the significant contribution of COMT in modulating the dynamics of dopamine overflow in the PFC. Moreover they may explain the outcome of numerous population studies where genetic variation of COMT activity and number of CNS diseases show some association, and underscore the therapeutic potential of manipulating COMT activity to alter dopaminergic neurotransmission in the PFC.

Catestatin (Human Chromogranin A352-372) Induces Lipolysis and Fatty Acid Oxidation through Regulation of Adrenergic and Leptin Signaling

Sushil K. Mahata[1,2], **Christine U. Vu**[1], **Stefano Gentile**[1,3], **Gautam K. Bandyopadhyay**[1] **and Daniel T. O'Connor**[1,2]

[1]*VA San Diego Healthcare System San Diego, California, USA;* [2]*University of California, San Diego La Jolla, California, USA;* [3]*University of Calabria Cosenza Italy*

BACKGROUND: The secretory proprotein Chromogranin A (CHGA in humans, Chga in mice) gives rise to several peptides of biological importance, which include the dysglycemic hormone pancreastatin (PST: CHGA250-301), the vasodilator vasostatin (CHGA1-76), and the antihypertensive, antiadrenergic, cardiosuppressive and angiogenic peptide catestatin (CST: CHGA352-372). The increased adiposity in hyperadrenergic, hyperleptinemic and insulin-sensitive Chga knockout (Chga-KO) mice is due to resistance to catecholamines and leptin.

OBJECTIVES: Since catestatin inhibits catecholamine secretion and catecholamines inhibit leptin secretion, we reasoned that CST would reduce obesity by restoring adrenergic receptor and leptin receptor sensitivity through normalization of catecholamine and leptin levels.

METHODS AND RESULTS: Chronic CST administration (5 µg/g BW/day for 16 days) to Chga-KO mice reduced epididymal fat pad size to WT level and decreased plasma triglyceride levels. This decrease in overall lipid content is caused in part by increased lipolysis as evident by increased glycerol and non-esterified fatty acid levels in plasma. CST treatment of Chga-KO mice also lowered plasma leptin to a level below WT and restored leptin action by reversing the desensitization effect of chronic leptin excess. Since adipose tissue in CST-treated mice showed increased palmitate oxidation but decreased incorporation into lipids, we believe that CST inhibits the expansion of adipose tissue and also promotes fatty acid uptake in liver for oxidation. In liver, CST augmented the expression of acyl-CoA oxidase 1 (Acox1), carnitine palmitoyltransferase 1a (Cpt1a), uncoupling protein 2 (Ucp2), and peroxisome proliferator-activated receptor-alpha (Pparα) genes involved in fatty acid oxidation. In contrast, CST had no effect on the expression of lipogenic genes such as sterol regulatory element-binding protein 1 (Srebp-1) and peroxisome proliferator-activated receptor-gamma (Pparγ Interestingly, CST stimulated the expression of cluster of differentiation 36 (Cd36), a transporter mediating cellular uptake of long-chain fatty acids, as well as the lipogenic gene Gpat4 (glycerol-3-phosphate acyltransferase-4). This indicates that CST stimulates fatty acid incorporation into triglycerides but not de novo lipogenesis. Overall, CST appears to promote lipid flux from adipose tissue toward liver for catabolism. CST reversed the hyperleptinemia of Chga-KO mice and improved leptin signaling as determined by phosphorylation of AMPK and Stat3. Both the α-antagonist phentolamine and CST potentiated the lipolytic effects of b-AR agonist iso-proterenol in primary adipocytes. These findings suggest that CST recapitulates the lipolytic effect of the a-AR antagonist phentolamine. This commonality of CST with phentolamine, coupled with its ability to inhibit the action of a-AR agonist phenylephrine, suggests that CST acts by suppressing α-AR signaling. To further clarify the interactions between CST and leptin pathways and to establish CST as an anti-obesity factor, we examined the effects of CST in leptin resistant high fat diet-induced obese (DIO) mice and leptin deficient Ob/Ob mice. CST treatment reduced body weight and adipose mass in DIO mice without reducing food intake. In contrast, CST decreased (by >25%) food intake in Ob/Ob mice. Interestingly, CST enhanced leptin effects on adipose tissue metabolism and signaling in both DIO and leptin-deficient Ob/Ob mice.

CONCLUSIONS: We conclude that CST promotes lipolysis and fatty acid oxidation by blocking α-adrenergic receptor signaling as well as by enhancing leptin receptor signaling. Therefore, we propose CST as a promising candidate for anti-obesity.

Triple Deuteration Modifies Pharmacokinetics and Pharmacodynamics of Dopamine in PD Models

Frank Schneider[1], Rudolf-Giesbert Alken[1], Torun Malmlöf[2], Björn Schilström[2] and Torgny Svensson[2]

[1]CDRD Berolina AB, Svedala, Sweden; [2]Karolinska Institutet, Stockholm, Sweden

BACKGROUND: L-DOPA, active, or toxic metabolites might contribute to motor and non-motor complications e.g. dyskinesia. Replacement of hydrogen by deuterium might reduce the metabolic breakdown rate of the active substance and the formation of adverse metabolites.

OBJECTIVES: These studies investigate the effects of deuterated L-DOPA + (+ carbidopa) on striatal DA availability and metabolic degradation and the corresponding motor and side effects.

METHODS AND RESULTS: The $\alpha\beta\beta$-triple deuterated BDD-20602 + did not show different peripheral plasma concentration kinetics in rats but generates significantly elevated and sustained striatal DA concentrations and a decreased DOPAC/DA ratio in comparison to L-DOPA +. The increase in striatal norepinephrine (NE) after L-DOPA + is nearly abolished after administration of BDD-20602 +. The equivalent single dose of BDD-20602 + in improvement of motor function was determined to be 60% of L-DOPA + by investigating the dose-response relation of contra-lateral rotations in the 6-OH-DA rat model. This was confirmed by comparing the effects of chronic treatment with 8 mg/kg L-DOPA, 8 mg/kg BDD-20602 and 5 mg/kg BDD-20602 in the cylinder test. BDD-20602 was more effective than L-DOPA at the same dose but did not induce more dyskinesia (abnormal involuntary movements). Furthermore 5 mg/kg BDD-20602 + showed the same motor effect and induced significantly less dyskinesia as 8 mg/kg L-DOPA +. BDD-20602 + (5 mg/kg) did not change the DA/NE ratio in PFC whereas the same L-DOPA + dose doubled the ratio.

CONCLUSIONS: Triple deuteration of L-DOPA reduces the metabolic degradation of the corresponding DA by MAO and DBH and increases the striatal availability. A lower DOPAC/DA ratio indicates that the highly toxic intermediate metabolite DOPAL is diminished as well. Reduced induction of dyskinesia at the same motor effect doses results in a broadening of the therapeutic window. BDD-20602 might open the door for a tuned L-DOPA alternative optimized for unmet needs in PD.

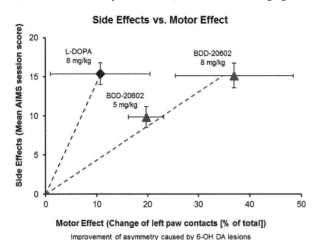

Side Effects vs. Motor Effect

y-axis: Side Effects (Mean AIMS session score)

x-axis: Motor Effect (Change of left paw contacts [% of total])

Improvement of asymmetry caused by 6-OH DA lesions

Understanding Catecholamine Metabolism as a Guide for Assessment of Sympathoadrenal Function and Dysfunction in Health and Disease

Graeme Eisenhofer

Institute of Clinical Chemistry and Laboratory Medicine and Department of Medicine, University Hospital of Dresden, Dresden, Germany.

The concept that most norepinephrine turnover occurs within sympathetic nerves due to leakage of the transmitter from vesicular stores was proposed in the 1960s by IJ Kopin. Development of assays to measure 3,4-dihydroxyphenylglycol, the principal deaminated metabolite of norepinephrine, established the almost exclusive neuronal origins of the metabolite. Related studies involving blockade of neuronal and vesicular monoamine transporters and comparisons of metabolism of endogenously released with exogenously administered radiolabeled catecholamines confirmed that at rest over 85% of the norepinephrine produced in sympathetic nerves is metabolized intraneuronally by monoamine oxidase and that at least 80% of this is due to leakage of norepinephrine from vesicular stores with the remainder reflecting metabolism after reuptake. Development of assays to measure the metanephrines, metabolites produced by catechol-O-methyltransferase, established that O-methylation represents a minor pathway of catecholamine metabolism. Studies involving sampling of blood flowing into and out of various organs showed that the single largest source of circulating metanephrines are the adrenals where the metabolites are produced from catecholamines leaking from vesicular stores. These studies also indicated substantial production and metabolism of NE and dopamine within mesenteric organs, where sulfate-conjugation plays an important metabolic role; the liver was confirmed as the primary site for formation of vanillylmandelic acid. The above findings have important implications for correct understanding of the involvement of catecholamine systems in health and disease. The contribution of norepinephrine leakage to turnover provides a crucial mechanism for "gearing down" the requirement for increases in catecholamine synthesis to match increases in release, thereby providing sympathetic nerves with an extended capacity for sustainable release rates than otherwise possible. This leakage, however, also has implications for neuropathological processes involving production of toxic deaminated aldehyde metabolites. Development of improved methods for diagnosis of pheochromocytoma reflects another advance that followed from improved understanding of catecholamine metabolism.

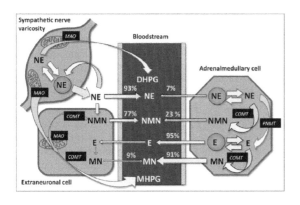

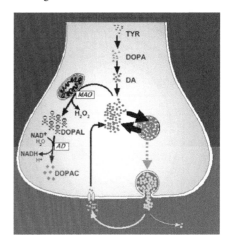

Multiplex Immunoassay for Catecholamines

Nadja Baur[1], Nicole Schneiderhan-Marra[1], Esther Booltink[2], Georg Manz[2], Thomas Joos[1] and Bernhard Manz[2]

[1]Natural and Medical Sciences Institute at the University of Tübingen, Reutlingen, Germany; [2]Labor Diagnostika Nord GmbH & Co. KG, Nordhorn, Germany

BACKGROUND: In an aging society, there is an increasing need for the diagnosis of hypertension, degenerative heart diseases and endocrine tumors like the pheochromocytoma. Determination of plasma and urinary biogenic amines like catecholamines and metanephrines is generally considered as a principal test for the clinical chemical diagnosis of these diseases. Currently, the determination of these analytes is performed using chromatographic techniques such as HPLC or ELISA methods. However, all chromatographic methods are complicated and cost-intensive and ELISA methods are labor-intensive and time-consuming. Therefore, there is great interest to measure these analytes in an efficient and cost-effective multiplex format.

OBJECTIVES: Using bead-based technology, we developed an assay that allows the simultaneous and quantitative determination of adrenaline, noradrenaline and dopamine in urine.

METHODS AND RESULTS: Urinary catecholamines were extracted using a cis-diol-specific affinity gel. Biotinylation was performed using amine-reactive NHS ester chemistry, followed by an enzymatic O-methylation. The extract containing the modified catecholamines was incubated with spectrally distinguishable magnetic bead sets, each carrying a specific antibody for adrenaline, noradrenaline or dopamine. Modified catecholamines bound to the specific antibodies were visualized via Streptavidin-R-Phycoerythrin. The read-out was performed using the Luminex FLEXMAP 3D system (Luminex Corp., Austin, TX, USA).

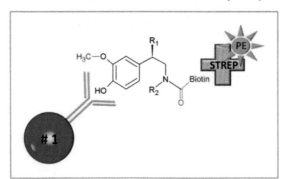

Using this 3-plex bead-based immunoassay we were able to simultaneously and quantitatively determine adrenaline, noradrenaline and dopamine in urine samples.

CONCLUSIONS: This bead-based immunoassay allows the simultaneous and quantitative determination of adrenaline, noradrenaline and dopamine in urine in a single assay. The analytical procedures are simple, cost-effective, time- and reagent-saving. We also have preliminary results for a 3-plex bead-based immunoassay for the quantitative determination of serotonin, metanephrine and normetanephrine in urine samples.

Assay characteristics	Adrenaline	Noradrenaline	Dopamine
LOD	0.18 ng/mL	0.53 ng/mL	1.78 ng/mL
LLOQ	0.47 ng/mL	2.34 ng/mL	2.34 ng/mL
Intra-assay precision	12 %	7 %	10 %
Inter-assay precision	10 – 18 %	7 – 18 %	4 – 18 %
Recovery	84 – 94 %	84 – 96 %	89 – 108 %
Cross-reactivity	no cross-reactivity could be observed		

Modifiers of Neurofibromatosis-Associated Tumors and the Potential Role for Catecholamines and Sex Differences

Min-Hyung Lee[1], Georgette Jones[1], Jessica Amlin-Van Schaick[1], Christina DiFabio[1], Melissa Whitman[1], Uma Shankavarum[2], Thanh Huynh[3], Sungjin Kim[4], Karel Pacak[3], Karl W. Broman[4] and Karlyne M. Reilly[1]

[1]*MCGP, NCI, FNLCR, NIH, Frederick, MD;* [2]*ROB, NCI, NIH, Bethesda, MD;* [3]*NICHHD, NIH, Bethesda, MD;* [4]*Dept. of Biostatistics and Medical Informatics, School of Medicine and Public Health, Univ. of Wisconsin, Madison, WI.*

Neurofibromatosis type 1 (NF1) is one of the most common genetic diseases of the nervous system, affecting 1 in 3500 people. NF1 patients are susceptible to astrocytoma in the central nervous system and pheochromocytoma in the adrenal gland, among other tumor types. A study of monozygotic twins and more distant relatives with NF1 has demonstrated that modifier genes unlinked to the *NF1* gene affect the susceptibility to different aspects of the disease. We are interested in understanding what "normal" physiological pathways affect the risk of developing cancer, and whether an understanding of this underlying biology can be used to predict who in the population is most at risk for developing particular cancers. We are particularly interested in how the differences in normal male and female physiology could lead to differences in cancer risk.

We are using the *Nf1 −/+ ;Trp53 −/+ cis* mouse model of NF1 to map modifiers of NF1-associated tumors in an unbiased genetic screen. We have identified a male-specific modifier of astrocytoma, *Arlm1*, on mouse

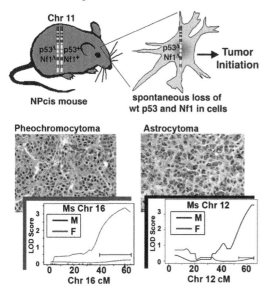

Chr 12, and a female-specific modifier of pheochromocytoma, *Pheom1*, on mouse Chr 16. Through cross-species bioinformatic analysis we have identified *Cdca7l*, encoding the R1 transcription factor, as a top candidate for the *Arlm1* locus. R1 is a co-transcription factor with Myc, and may play a role in cancer through effects on cell proliferation and apoptosis downstream of Myc. Alternatively, R1 plays a role in normal physiology by inhibiting Mao and regulating downstream catecholamines. In a screen of the LOPAC1250 library of pharmacologically active compounds, we found that compounds active in the serotonin and dopamine pathways affected growth of astrocytoma cells. These findings raise interesting questions about the role of catecholamine metabolism in NF1-associated tumorigenesis, and whether differences in the catecholamine axis between males and females accounts for the sex-specificity of modifiers of astrocytoma and pheochromocytoma.

Catecholamine Metabolism in Chromogranins Knock-Out Mice

Marta R. Pardo[1]**, Daniel Pereda**[1]**, José-David Machado**[1]**, María-Rosa Arnau**[1]
and Ricardo Borges[1,2]

[1]Pharmacology Unit, Medical School; [2]IUBO-AG, La Laguna University, Tenerife, Spain

BACKGROUND: We have demonstrated that the catecholamine (CA) storage in chromaffin vesicles strongly depends on their chromogranins (Cgs) content and that the cytoplasmatic levels of CA in chromaffin cells were reduced when Cgs were absent. Amperometric data have showed the impaired of CA uptake in secretory vesicles in Cgs-KO mice and they are unable to accumulate more amines (Díaz-Vera, 2012).

OBJECTIVES: We have studied their metabolism of CA in adrenal medulla, their urinary excretion and their distribution in chromaffin vesicles. In addition, CgA&B-KO mouse has never been characterized and its metabolism is unknown.

METHODS AND RESULTS: Mice exhibited overweight and glucose resistance. The lack of Cgs is largely disturbed in isolated granules distribution in density gradients showing a greater CA:ATP ratio. The urinary excretion of norepinephrine, DHPG and dopamine in these mice is 1.6 times more than controls in 24-h collection. Values are normalized by creatinine measured by Jaffe's reaction.

The TH expression is 2-fold greater than controls, measured by western blot, but there is no difference in the production rate of L-DOPA from TH.

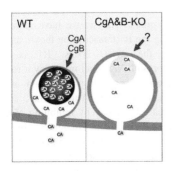

Neither MAOA nor B activity change in Cgs-KO mice, measured by a one-step fluorometric method.

CONCLUSIONS: Our results show that Cgs-KO mice are able to synthesize more CA in a stressful situation as TH expression is increased. Conversely, as CA metabolism seems not to be impaired and urinary excretion of CA is increased in Cgs null mice, we conclude that leak-out of amine from the cytosol of chromaffin cells is of importance.

Supported by MICINN (BFU2010-15822) and CONSOLIDER (CSD2008-00005) and ACIISI/FEDER PI2007/017&C2008/01000239.

Translation Research in Autism and Cancer: Novel Roles of MAO

Jean C. Shih[1], Marco Bortolato[1], Boyang Jason Wu[2] and Kevin Chen[1]

[1]University of Southern California, Los Angeles, California, USA;
[2]Cedars-Sinai Medical Center, Los Angeles, California, USA

BACKGROUD: Monoamine oxidase A and B (MAO A and B) are mitochondrial isoenzymes, degrade neurotransmitters, serotonin (5-hydroxytryptamine, 5-HT), dopamine (DA) and norepinephrine (NE) and produce hydrogen peroxide. MAO A knockout (KO) mice exhibit high brain levels of 5-HT, DA and NE, and antisocial traits and aggression. MAO A/B KO mice, the brain levels of monoamines and severity of behavioral alterations is significantly greater than those in MAO A KO mice.

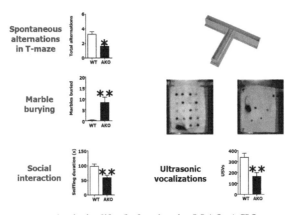

Autistic-like behavior in MAO A KO

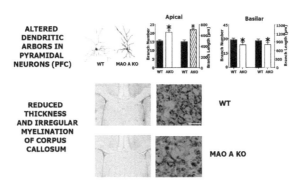

Autistic-like morphology in MAO A KO

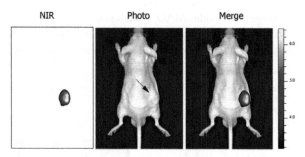

MHI148-clorgyline

NIR	Photo	Merge

Localization of MHI148-clorgyline in tumor

OBJECTIVES: To use these KO mice for developing drugs for disorders associated with MAO.

RESULTS AND CONCLUSIONS: MAOA and A/B KO mice exhibit numerous behavioral hallmarks and morphological changes of autism-spectrum disorder (ASD), and that P1 to P7 is the most critical period to rescue the abnormal behavior by serotonin or ketanserin, a serotonin-2A receptor antagonist.

Also interesting, MAO A KO in both cancer cells and host reduced the progression and metastasis of prostate cancer. Translating this basic information to clinical suggested that MAO A inhibitors may be used for the treatment of prostate cancer.

This conjugate delivers the clorgyline (MAO A inhibitor) specifically to tumors and can be visualized by NIR imaging.

MHI148-clorgyline was injected to nude mice and imaged.

Supported by NIMH Merit Award, NIH RO1 MH39085, Boyd and Elsie Welin Professorship.

COMT is a Major Contributor to Pain Perception: from Humans to Rats and Back to Humans

Luda Diatchenko[1]

[1]Center for Neurosensory Disorders, University of North Carolina, Chapel Hill, NC, USA.

BACKGROUND: Catechol-O-methyltransferase (COMT) is an enzyme responsible for degrading catecholamines and thus represents a critical component of homeostasis maintenance, including contribution to cognition, stress response and pain perception.

OBJECTIVES: I review a serial of recent studies linked COMT to pain perception.

METHODS AND RESULTS: First, in a human association studies, three major haplotypes of *COMT*, designated as low pain sensitive (LPS), average pain sensitive (APS), and high pain sensitive (HPS) have been identified based on a carrier's response to experimental pain stimuli. These three haplotypes account for 11% of the variability to experimental pain sensitivity in young women and are predictive of the risk of onset of a common musculoskeletal facial pain disorder. The LPS haplotype produces higher levels of COMT enzymatic activity than the APS or HPS haplotypes. Next, the pharmacological inhibition of COMT in rats results in mechanical and thermal hypersensitivity that is reversed by the nonselective β-adrenergic antagonist propranolol, or by the combined administration of selective β_2- and β_3-adrenergic antagonists. In contract, the administration of β_1-adrenergic, α-adrenergic, or dopaminergic receptor antagonists fail to alter COMT-dependent pain sensitivity. These data provide the first direct evidence that low COMT activity leads to increased pain sensitivity via a $\beta_{2/3}$-adrenergic mechanism. Last, a double-blind, placebo-controlled, two-period crossover pilot study of efficacy of propranolol, non-selective β-blocker, in 40 female patients suffering from TMD was conducted. Propranolol significantly reduced a composite measure of clinical pain, and showed a trend towards decreasing experimental pressure and heat pain ratings compared to placebo. When stratified by the *COMT* high activity haplotype (LPS), a significant beneficial effect of propranolol on pain perception was noted in subjects not carrying this haplotype, a diminished benefit was observed in the heterozygotes, and no benefit was noted in the homozygotes.

CONCLUSIONS: A serial of recent studies linked COMT to pain perception, identified genetic variation in *COMT* gene correlating with human secitivity to noxios stimuli, and revealed new drug targets that then has been be translated into the pharmacological treatment of common pain conditions. Human *COMT* haplotypes may serve as genetic predictors of treatment outcomes and permit the identification of a subgroup of patients who will benefit from propranolol therapy.

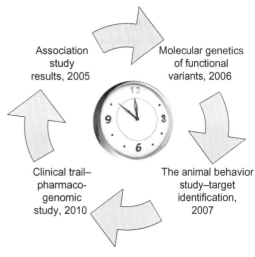

Association study results, 2005

Molecular genetics of functional variants, 2006

Clinical trail– pharmaco-genomic study, 2010

The animal behavior study–target identification, 2007

FIGURE 1

Translational clock – closing the circle.

Novel COMT Inhibitors in Parkinson Disease

Patricio Soares-da-Silva

*R&D, BIAL—Portela & Cª, S.A., 4745-457 S. Mamede do Coronado, Portugal. Dept. Pharmacology &
Therapeutics, Faculty of Medicine, University Porto, Porto, Portugal*

Despite metabolic instability and arguable safety profile, levodopa (L-DOPA) remains a landmark in the symptomatic treatment of Parkinson's disease (PD). PD is a chronic neurodegenerative disorder associated with a reduction of the levels of endogenous neurotransmitter dopamine in the striatum. Levodopa is the biological precursor of dopamine, which is able to modulate cerebral levels of dopamine by penetrating into the brain. Clinical efficacy of the therapy can dramatically be improved by blocking the metabolic deactivation of levodopa in peripheral tissues. One of the principal metabolizing enzymes involved in the breakdown of levodopa is catechol-O-methyltransferase (COMT). COMT inhibitors help to sustain the continuous delivery of dopamine to the striatum and thereby motor-related symptoms of PD are diminished. The lecture will provide an overview of the major COMT inhibitors (entacapone, tolcapone and opicapone), summarizing the main pharmacokinetic and pharmacodynamic properties as well as highlighting their clinical relevance in the symptomatic treatment of PD. Opicapone made its way through an unique path of drug development in which molecular modelling provided the design of high potency, low toxicity, and long acting COMT inhibitors. These characteristics are expected to change the way COMT inhibitors are perceived for their clinical use, namely on efficacy, safety and posology. Opicapone's drug development is, at present, in progress through phase III clinical trials as once-daily dosing as adjunct to levodopa plus AADC therapy in PD patients with motor fluctuations. As a pharmacological tool, the availability of opicapone is a particular important resource, since it will allow for a better understanding of the inhibitor-enzyme complex, namely with regards to the functional relevance of low dissociation rate constant (koff) of the inhibitor from the enzyme-inhibitor complex as key determinant in defining opicapone in vivo pharmacological profile and clinical use in PD.

Effect of Opicapone, a New Catechol-*O*-Methyltransferase Inhibitor, in Levodopa Pharmakokinetics in the Cynomolgous Monkey

Maria João Bonifácio[1], Jane Suzanne Sutcliffe[2], Leonel Torrão[1], Lyndon Wright[1] and Patrício Soares-da-Silva[1,3]

[1]*Dept. Research & Development, BIAL — Portela & Cą, S.A., 4745—457 S. Mamede do Coronado, Portugal;*
[2]*Maccine Pte Ltd. 10 Science Park Road, #01—05 The Alpha, Singapore Science Park II, Singapore;*
[3]*Dept. Pharmacology & Therapeutics, Faculty of Medicine, University Porto, Porto, Portugal.*

BACKGROUND: Opicapone (1) is a novel peripheral catechol-*O*-methyltransferase (COMT; EC 2.1.1.6) inhibitor developed to be used as an adjunct to the levodopa/aromatic aminoacid decarboxylase inhibitor therapy of Parkinson's disease (PD).

OBJECTVES: This study evaluated the effect of opicapone on the systemic and central bioavailability of levodopa and related metabolites, when chronically administered to cynomolgous monkeys.

METHODS AND RESULTS: This was a crossover design study where six animals were randomized to two groups that received vehicle or 100 mg/kg opicapone for 14 days, with a washout of 5 days. Twenty-three hours after last administration animals were given levodopa/benserazide (12/3 mg/kg). Levodopa and related metabolites was determined in dialysates collected from microdialysis probes placed in the substantia nigra, dorsal striatum and prefrontal cortex, and also in plasma. The activity of COMT was determined in erythrocytes. The daily administration of 100 mg/kg opicapone over a period of 14 days resulted in significant increased levodopa systemic exposure (from 324,485 to 638,463 ng.h/ml) and reduced exposure to 3-OMD (from 548,056 to 111,675 ng.h/ml). These changes were accompanied by ~80—85% reduction in erythrocyte COMT activity. Opicapone levels in plasma attained their maximum at 1 h post-administration with a C_{max} (ng/ml) of 964 in males and 1170 in females; AUC (h.ng/ml) was 2319 in males and 1762 in females. Opicapone increased levodopa exposure (from 2,031 to 3,352 ng.h/ml) in the dorsal striatum dialysate and reduced exposure to 3-OMD (from 8,213 to 1,631 ng.h/ml). Similar findings were observed in substantia nigra and prefrontal cortex.

CONCLUSIONS: The marked increase in systemic and central levodopa bioavailavility obtained with opicapone is the result of the exceptionally prolonged COMT inhibition. Opicapone is therefore an excellent candidate to fulfill the unmet need for COMT inhibitors that can lead to more sustained levodopa levels in PD patients.

Opicapone

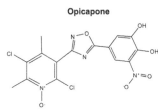

Plasma levodopa

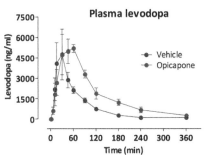

Reference

1. J Med Chem (2010) 53:3396—3411.

Does Impaired Aldehyde Detoxification Play a Role in Parkinson's Disease?

Randy Strong and Margaret Wey

Department of Pharmacology[1], Barshop Institute for Longevity and Aging Studies; [2], University of Texas Health Science Center at San Antonio, Geriatric Research, Education and Clinical Center, South Texas Veterans Health Care Network; [3], San Antonio, Texas

BACKGROUND: Parkinson's disease (PD) is the 2nd most common neurodegenerative disorder affecting 2 − 5 % of the elderly. Symptoms are primarily motor dysfunction as a result of loss of dopaminergic neurons in the substantia nigra (SN). Proposed causes of idiopathic PD pathogenesis include oxidative stress, mitochondrial dysfunction, and protein aggregation. Amongst sources of oxidative stress, biogenic aldehydes are particularly toxic due to their reactivity and relatively long half-life. Aldehydes are primarily detoxified by aldehyde dehydrogenases (ALDH) and ALDH1a1 and ALDH2 are known to be expressed in SN dopamine neurons. Previous studies have reported elevated levels of biogenic aldehydes in the brains of patients with PD. Reduced *ALDH1* expression in surviving midbrain dopamine neurons has been reported in PD brains. In addition, impaired complex I activity reduces the availability of the NAD^+ co-factor required by ALDH isoforms to catalyze removal of biogenic aldehydes.

OBJECTIVES: We hypothesized that chronically decreased function of multiple aldehyde dehydrogenases consequent to exposure to environmental toxins and/or reduced ALDH expression, plays an important role in the pathophysiology of PD.

METHODS AND RESULTS: Mice null for Aldh1a1 and Aldh2. $Aldh1a1^{-/-} \times Aldh2^{-/-}$ mice exhibited age-dependent deficits in motor performance assessed by gait analysis, performance on an accelerating rotarod, vertical pole test and challenging beam traversal. Intraperitoneal administration of L-DOPA alleviated motor deficit. We observed a significant loss of neurons immunoreactive for tyrosine hydroxylase (TH) in SN as well as a reduction of dopamine and metabolites in the striatum of $Aldh1a1^{-/-} \times Aldh2^{-/-}$ mice. We also observed significant increases in biogenic aldehydes, including 4-hydroxynonenal (4-HNE) and the aldehyde intermediate of dopamine metabolism, 3,4-dihydroxyphenylacetaldehyde (DOPAL).

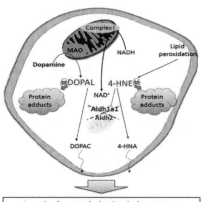

Impaired motor behavioral phenotypes

CONCLUSIONS: These results support the hypothesis that impaired detoxification of biogenic aldehydes may be important in the pathophysiology of PD and suggest that $Aldh1a1^{-/-} \times Aldh2^{-/-}$ mice may be a useful animal model of PD.

The Role of Aldehyde Detoxification in Parkinson's Disease

Margaret Chia-Ying Wey[1,2,3]**, Shou-Shu Wang**[3]**, Paul Anthony Martinez**[2]**, Vanessa Martinez**[2]**, Patricia Sullivan**[4]**, David Goldstein**[4]**, Elizabeth Fernandez**[1,2,3] **and Randy Strong**[1,2,3]

[1]*Department of Pharmacology;* [2]*Barshop Institute for Longevity and Aging Studies;* [3]*University of Texas Health Science Center at San Antonio, Geriatric Research, Education and Clinical Center, South Texas Veterans Health Care Network;* [4]*San Antonio, Texas and Institute of Neurological Disorders and Stroke, Bethesda, Maryland*

BACKGROUND: Parkinson's disease (PD) is an age-associated neurodegenerative disorder characterized by motor dysfunction resulting nigro-striatal neurodegeneration. Studies indicate that oxidative damage contributes to PD pathogenesis. Biogenic aldehydes are particularly toxic due to their reactivity and relatively long half-life. Accumulation of biogenic aldehydes in PD brains has been reported. Aldehyde dehydrogenases (ALDH) play a major role in brain aldehydes detoxification. We reported that mice null for two Aldh isoforms expressed in midbrain dopamine neurons, Aldh1a1 and Aldh2, show manifestations of PD, including loss of dopamine neurons and age-related motor deficits that are rescued by L-DOPA.

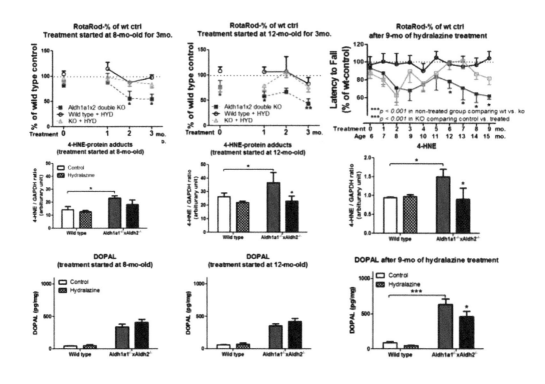

OBJECTIVES: We hypothesized that aldehyde-trapping agents may be cytoprotective in conditions of increased "aldehyde load", and aldehyde-trapping agent, hydralazine, may be beneficial in PD.

METHODS AND RESULTS: Hydralazine prevented 4-HNE-induced toxicity in PC12 cells. Hydralazine (250 mg/L) was delivered in drinking water for 3−9 months to $Aldh1a1^{-/-} \times Aldh2^{-/-}$ mice. Delivery of hydralazine not only attenuated motor deficits in $Aldh1a1^{-/-} \times Aldh2^{-/-}$ mice, it reduced midbrain 4-HNE levels, suggesting that aldehyde-trapping agent, hydralazine, can scavenge aldehydes *in vivo*. In contrast, the striatal DOPAL content in mice treated with hydralazine for 3 months was not different from those fed drinking water. However, 9-month hydralazine treatment $Aldh1a1^{-/-} \times Aldh2^{-/-}$ mice produced significant reductions in both midbrain 4-HNE and striatal DOPAL levels, as well as improved motor behavior.

CONCLUSIONS: Mice treated with hydralazine showed improved motor function. Three-month hydralazine treatment reduced midbrain 4-HNE but not striatal DOPAL level; 9-month hydralazine treatment, on the other hand, reduced both 4-HNE and DOPAL in the nigrostriatum. The results supported our hypothesis that the aldehyde-trapping agent, hydralazine, may provide a new therapeutic approach to PD As a currently FDA-approved drug, hydralazine may potentially be readily approved as a neuroprotective agent in PD.

Pharmacological Profile of Opicapone in Wistar rat

Maria João Bonifácio[1], Leonel Torrão[1], Ana Loureiro[1], Carlos Fernandes-Lopes[1], Lyndon C. Wright[1] and Patrício Soares-da-Silva[1,2]

[1]*Dept. Research & Development, BIAL — Portela & Ca, S.A., 4745-457 S. Mamede do Coronado, Portugal;*
[2]*Dept. Pharmacology & Therapeutics, Faculty of Medicine, University Porto, Porto, Portugal.*

BACKGROUND: Opicapone (1) was developed to fulfill the unmet need for safer COMT inhibitors in the treatment of Parkinson's disease, with an improved pharmacodynamic profile to provide sustained levodopa availability.

OBJECTIVES: This study aimed at characterizing some pharmacodynamic and pharmacokinetic properties of opicapone in Wistar rats.

METHODS AND RESULTS: COMT activity was determined in rat tissues and erythrocytes as previously described (2). Catecholamines were quantified by HPLC-ED and opicapone was measured by LC-MS/MS. Administration of opicapone (3 mg/kg, po) resulted in marked inhibition ($>80\%$ inhibition) of liver and kidney COMT activity from 1 to 8 h post-administration, with the enzyme slowly recovering 70–86% of control values at 48 h. Opicapone did not affect brain COMT activity. Maximum opicapone levels in plasma were achieved at 4 h post-administration with a C_{max} of 661.5 ng/ml; AUC was 2409.6 h.ng/ml. The oral administration of increasing doses of opicapone resulted in dose dependent inhibition of COMT activity in liver, kidney and erythrocytes, as depicted in table 1. Opicapone is equipotent in inhibiting peripheral COMT at 2 and 6 h post-administration. Administration of opicapone to rats dosed with 12 mg/kg levodopa plus 3 mg/kg benserazide resulted in sustained increase (at least up to 24 h) in levodopa plasma levels with 3-O-methyldopa (3-OMD) levels significantly decreased over the course of the experiment, from 2 h post-administration of opicapone. Significant increases in levodopa, DOPAC, dopamine and HVA and a significant reduction in 3-OMD levels were observed in the brain.

CONCLUSIONS: Opicapone is a novel, very potent, peripheral COMT inhibitor endowed with a long-acting inhibitory profile, that when used as an adjunct to levodopa/benserazide administration leads to a sustained increase in peripheral and central levodopa bioavailability with concomitant reduction of levodopa methylation.

Table 1 ED_{50} Values (in mg/kg) with Respective 95% Confidence Intervals

	2 h	6 h
Liver	0.7(0.6; 1.0)	1.1 (1.0; 1.3)
Kidney	0.4(0.3; 0.5)	0.5 (0.4; 0.7)
Erythrocytes	0.9(0.7; 1.3)	1.4 (1.2; 1.6)

References

1. J Med Chem (2010) 53:3396–3411.
2. Mol Pharmacol (2002) 62:795–805.

Catecholamine Receptors and Catecholaminergic Signaling

Lee Eiden and David Goldstein

Perhaps the most important advances in catecholamine research in the last ten years has been the elucidation of the structures of the G-protein coupled receptors with which the catecholamines interact as first messengers, and the understanding that these receptors, perhaps better called 'seven transmembrane receptors', initiate signaling through the arrestin protein family that may exceed the scope of G-protein coupled signaling that these receptors 'classically' initiate. It was highly appropriate that discussion of the mechanisms of signal transduction revealed by analysis of beta receptor-ligand crystals led by Brian Kobilka, and a cogent history of the GPCRs leading to the modern conceptualization of translationally and clinically relevant 'biased ligands' for the multiple G-protein dependent and independent signaling pathways by plenary speaker Robert Lefkowitz, was followed within a few weeks of the meeting by the award of the Nobel Prize in Chemistry to both men.

The structural bases for both receptor selectivity, and biased signaling, was explored in great detail in sessions on catecholamine structure-function studies, the functional significance of catecholamine receptor oligomerization, and molecules involved in receptor trafficking, recycling and degradation, and their physiological and pathophysiological regulation in CNS neurons. In brief, the 'state-of-play' of a given receptor within the cell, including its association with molecules other than the first messenger, and its abundance and localization within the cell, contribute far more to the dynamics of neurotransmission than has been classically envisioned. Each of the interactions leading to altered receptor display, ligand recognition, and downstream signaling capability are of course potential avenues for understanding the actions of existing drugs, including drugs of abuse (see Theme G), and the development of new ones.

Neuroimaging of receptors received considerable attention, as this is a rapidly expanding observational link between insights gained by reductionist cell and molecular biological studies, the impact of pharmacological perturbation of these processes on human cognition and behavior, and neurological and psychological disease (see Themes E, F and G). A truly impressive array of 18F-labeled receptor ligands for high-resolution PET analysis is now available, and these are useful not only for imaging receptor abundance in disease, but the degree of receptor occupancy by endogenous catecholamine ligands under various conditions (e.g. cocaine use) using in vivo dynamic equilibrium binding to specific receptor subtypes in dorsal and ventral striatum, as well as limbic cortical regions such as hippocampus and amygdala.

The final sessions in Theme D were concerned with functional selectivity of receptor signaling, and exploiting this new information in the search for new neuropharmacological agents with greater selectivity for treatment of psychological illness including drug addiction, post-traumatic stress disorder, and mood and cognitive disorders. The actual signaling pathways activated by both G-protein coupling, and arrestin activation have been explored with ingenious experiments combining cell culture analyses with photoactivatable and photoactive sensors, and genetic augmentation, deletion and gene swapping, and their effects on appetitive behaviors, arousal, mood, and cognitive and motor function. Novel methodologies for revealing post-receptor activation signaling in vivo have begotten novel approaches for drug discovery. The final session on this topic, chaired by David Sibley, featured new techniques for high-throughput matching of GPCR subtypes and function to GPCR-associated ligand screening, based on both isosteric and allosteric drug-receptor interactions. The last presentation featured the characterization of a novel cAMP sensor linking GPCR ligand occupancy to activation of the MAP kinase ERK, a key signaing molecule for memory, learning and synaptic plasticity, demonstrating that the pleiotropy of GPCR signaling, for which the catecholamine receptors are paradigmatic, still remains incompletely plumbed by pharmacologists and cell biologists alike.

How Drugs Bind and Control their Targets: Characterizing GPCR Signaling through Long-Timescale Simulation

Ron Dror

Shaw Research, USA

Roughly one-third of all drugs act by binding to G-protein-coupled receptors (GPCRs) and either triggering or preventing receptor activation, but the process by which they do so has proven difficult to determine using either experimental or computational approaches. We recently completed a special-purpose machine, named Anton, that accelerates molecular dynamics simulations of biomolecular systems by orders of magnitude, enabling all-atom protein simulations as long as a millisecond (Science 330:341−6, 2010). Anton has made possible unbiased simulations in which drugs spontaneously associate with GPCRs to achieve bound conformations that match crystal structures almost perfectly (PNAS 108:13118−23, 2011; Nature 482:552−6, 2012). Simulations on Anton have also captured transitions of the β2-adrenergic receptor, a prototypical GPCR, between its active and inactive states, allowing us to characterize the mechanism of receptor activation (Nature 469:236−40, 2011; PNAS 108:18684−9, 2011). Our results, together with complementary experimental data, suggest opportunities for the design of drugs that achieve greater specificity and control receptor signaling more precisely.

Structural basis of Subtype Selectivity in Dopamine D3 and D2 Receptors

Lei Shi

Weill Medical College of Cornell University, USA

All current antipsychotic drugs block dopamine D2-like receptors (D2R, D3R, and D4R) non-selectively, but it is generally thought that antagonism of D2R is essential for therapeutic efficacy. On the other hand, antagonists of the cognate D3R have been proposed as potential medications for the treatment of drug abuse.However, side effects can be a serious issue for the drugs targeting D2-like receptors due to non-selective blockade of the receptors and thereby their downstream effectors. In spite of the great importance of the clinical indications for drugs targeting the D2-like receptors, the full therapeutic potential of these targets has been difficult to explore rigorously due to the lack of drugs that are sufficiently selective. Substantial progress has been made in the discovery of D3R- or D2R-selective ligands, but the development of agents with appropriate physicochemical properties and pharmacokinetics for in vivo selectivity continues to challenge clinical translation. The recently described crystal structure of D3R should make progress possible as indicated by the initial structure analysis [1]. The D3R structure reveals an orthosteric binding site (OBS) located in the upper-half of the transmembrane domain containing the nonselective antagonist eticlopride. Remarkably, of the 18 eticlopride contact residues in D3R, 17 are identical in D2R, and one is similar [1]. This high degree of sequence identity in the OBS exposes the challenge in creating selective agents with drug-like properties, but also offers the first clues to the manner in which leads will have to be optimized to retain selectivity. Our recent analysis of a series of D3R-selective bitopic 4-phenylpiperazines has revealed important leads towards exploiting the structural basis of D3R over D2R selectivity. A key element is the identification of a divergent second binding pocket (SBP) that accommodates the arylamide moiety of the compounds. We also found that pharmacological efficacy depends on the binding mode of the 4-phenylpiperazine in the OBS, which is modulated by phenyl ring substitutions and linker length2. These findings are generalizable to other GPCRs in which the SBP can be targeted by bitopic or allosteric ligands.

References

1. Chien, E. Y.; Liu, W.; Zhao, Q.; Katritch, V.; Han, G. W.; Hanson, M. A.; Shi, L.; Newman, A. H.; Javitch, J. A.; Cherezov, V.; Stevens, R. C. Structure of the human dopamine D3 receptor in complex with a D2/D3 selective antagonist. Science 2010, 330, 1091−5.
2. Newman, A. H.; Beuming, T.; Banala, A. K.; Donthamsetti, P.; Pongetti, K.; Labounty, A.; Levy, B.; Cao, J.; Michino, M.; Luedtke, R. R.; Javitch, J. A.; Shi, L. Molecular Determinants of Selectivity and Efficacy at the Dopamine D3 Receptor. J Med Chem 2012.

Hormone-Mediated Activation of G Proteins by GPCRs: Insights from the Crystal Structure of a GPCR-G Protein Complex

Roger Sunahara

University of Michigan Medical School, USA

Recent advances in the structural biology of G protein-coupled receptors have helped to unravel the intricacies of ligand binding. Similarly structural and biochemical analyses of heterotrimeric G proteins have affirmed our understanding of the mechanism underlying effector interactions and GTPase activity. We recently elucidated the crystal structure of a prototypic GPCR, the b2-adrenergic receptor (b2AR), in a complex with the stimulatory G protein, Gs, trapped in its nucleotide-free state. These data have helped to delineate how hormone binding to GPCRs leads to GDP release on G proteins, the principle step that precedes in GTP binding and G protein activation. The crystal structure, together with data from single particle reconstructions by electron microscopy and deuterium exchange mass spectrometry, reveal dramatic changes in the G protein a-subunit. The crystal structure also suggests that G proteins may allosterically regulate the receptor by stabilizing a closed conformation on the extracellular face of the receptor. Consistent with these changes radioligand binding analyses suggest that G protein coupling slows ligand dissociation. These structural changes would account for the slower observed ligand dissociation rates and likely account for G protein-dependent high affinity agonist binding. Cumulatively these data support a model for the mechanism for receptor-mediated nucleotide exchange, G protein activation and agonist binding.

Towards Better Understanding of G(s) Coupling in Catecholamine Receptors

Hideaki Yano[1], Davide Provasi[2], Marta Filizola[2] and Jonathan Javitch[1]

[1]Columbia University, New York Psychiatric Institute; [2]Mount Sinai School of Medicine

BACKGROUND: Our understanding of conformational changes associated with receptor activation has been advanced profoundly by the recent crystal structure of agonist-bound β2 adrenergic receptor (β2AR) coupled to G(s). The large changes seen in G protein structure are thought to be related to receptor activation and GDP/GTP exchange in the G protein. The availability of a single snapshot with a single high affinity agonist leads to uncertainty regarding the extent to which conformational changes in G protein are conserved across different receptors as well as across other G proteins.

OBJECTIVES: We set out to understand the movement of the G(s) heterotrimer after activation of β2AR. Using bioluminescence resonance energy transfer (BRET), we made observations on relative movement within the G protein and between the receptor and the G protein. We compared the data to the movement mechanism proposed in the recent studies. Next we investigated if the conformational change of G(s) protein activation is different in another catecholamine receptor, the dopamine receptor D1 (D1R). Lastly, we studied G(olf) by the same approach and compared its conformational change with G(s) in the hope of understanding its activation mechanism.

METHODS AND RESULTS: We constructed biosensor G(s) with either luciferase or fluorescent protein inserted at ten distinct positions throughout the structure, avoiding the critical GTPase domain. Activation-induced changes in BRET report on the relative proximity and orientation between the donor and acceptor, which are fused in different strategic positions within the G protein subunits and the receptor. Many of the signature changes suggested in the crystal structure and electron microscopy studies including hinge movement of the helical domain are consistent with the BRET changes observed in G(s) coupled β2AR at the receptor to G alpha interface or the G alpha to beta-gamma subunit interface.

Using the same set of G(s) biosensors, D1R activation mediated change was studied. Remarkably, some of the movement within the G(s) heterotrimer is not as pronounced as with the β2AR suggesting that the conformational changes induced in the same G proteins may differ depending upon the activated receptor. Finally, taking advantage of high homology in amino acids, G(olf) biosensor constructs were made at the same discrete ten positions used for G(s). Interestingly, these biosensors did not show much of G protein conformational change.

CONCLUSIONS: Our extensive analysis of G(s) biosensors sheds light on potential activation differences within the G(s) heterotrimeric complex between β2AR and D1R. These results are now being explored in a structural context taking advantage of the crystal structures and cryomicroscopy studies of activated G protein. Comparison between G(s) and G(olf) gives us a perspective on possible differences in conformational movements despite the high sequence homology. These biosensors also can be used as an assay for screening drugs that selectively activates G(s) or G(olf) pathway.

Supported by NIH grants DA022413, and MH54137, and Lieber Center for Schizophrenia Research.

Receptor Heteromerization: Understanding the Functional and Pathological Role of the Dopamine D4 Receptor

Sergi Ferre

National Institute on Drug Abuse, IRP, NIH, DHHS, USA

Recent studies showing the ability of dopamine D4 receptors to heteromerize with other G protein-coupled receptors are shedding light on their functional and also pathological role in the central nervous system. D4 receptors are highly expressed in the pineal gland during the dark cycle, following a circadian rhythm. We have recently shown that the production of both melatonin and serotonin by the pineal gland is regulated by a circadian-related heteromerization of adrenergic and dopamine D4 receptors. Through a1B-D4 and b1-D4 receptor heteromerization dopamine inhibits adrenergic receptor signaling and blocks the synthesis of melatonin induced by adrenergic receptor ligands. These data provide a new perspective on dopamine function mediated by D4 receptors and constitute the first example of a circadian-controlled receptor heteromerization. Polymorphic variants of the dopamine D4 receptor have been consistently associated with attention-deficit hyperactivity disorder (ADHD). However, the functional significance of the risk polymorphism (variable number of tandem repeats in exon 3) is still unclear. We have recently shown that whereas the most frequent 4-repeat (D4.4) and the 2-repeat (D4.2) variants form functional heteromers with the short isoform of the dopamine D2 receptor (D2S), the 7-repeat risk allele (D4.7) does not. In the striatum, D4 receptors are localized in corticostriatal glutamatergic terminals, where they selectively modulate glutamatergic neurotransmission by interacting with D2S receptors. We have therefore postulated that dysfunctional D2S-D4.7 heteromers may impair presynaptic dopaminergic control of corticostriatal glutamatergic neurotransmission and explain functional deficits associated with ADHD.

Trace Amines and Their Receptors: New Opportunities for Modulation of Brain Dopaminergic Functions

Tatyana D. Sotnikova, Stefano Espinoza and Raul R. Gainetdinov

Istituto Italiano di Tecnologia, Genova, Italy

BACKGROUND: Structurally related to classical monoaminergic neurotransmitters, trace amines (such as beta-phenylethylamine, tyramine, octopamine and tryptamine) are found at low concentrations in the mammalian brain. While their physiological roles in mammals are generally unknown, the recently discovered group of G protein-coupled receptors, named trace amine associated receptors (TAARs), provided opportunity to explore their functions. The trace amine associated receptor 1 (TAAR1), which is in part associated with the monoaminergic neuronal circuitry, is the best characterized of the class, though still little is known about its regulation and function. TAAR1 can be activated not only by trace amines but also by a variety of monoaminergic compounds including amphetamines and monoamine metabolites.

OBJECTIVES: To investigate the physiological functions mediated by TAAR1 we performed series of experiments by using both in vivo and in vitro approaches.

METHODS AND RESULTS: By applying various experimental paradigms aimed to model dopaminergic dysregulation in mice lacking TAAR1 we investigated the potential role of TAAR1 in modulating dopamine-related functions such as movement control. Furthermore, we investigated the biochemical mechanism of interaction between TAAR1 and D2 dopamine receptors (D2R) and the role this interaction plays in D2R-related signaling and behaviors. In TAAR1 knockout (KO) mice we observed that TAAR1 generally exerts an inhibitory influence on the locomotion, so TAAR1 agonists inhibit dopamine-dependent locomotor activity, while effects of dopaminergic stimulation is enhanced in TAAR1 KO mice. Using a bioluminescence resonance energy transfer biosensor for cAMP in cellular assay, we demonstrated that the D2R antagonists haloperidol, raclopride, and amisulpride were able to enhance selectively a TAAR1-mediated increase of cAMP. Moreover, TAAR1 and D2R were able to form heterodimers when coexpressed in human embryonic kidney 293 cells, and this direct interaction was disrupted in the presence of haloperidol. In addition, in mice lacking TAAR1, haloperidol-induced striatal c-Fos expression and catalepsy were significantly reduced. These data indicate that TAAR1 and D2R have functional and physical interactions that could be critical for the modulation of the dopaminergic system by TAAR1 in vivo.

CONCLUSIONS: Taken together, these investigations suggest that TAAR1 may represent a novel target for the pharmacology of dopamine-related disorders such as schizophrenia, ADHD and Parkinson's disease. Other potential therapeutic applications of selective TAAR1 agonists and antagonists will be discussed. It is expected that further characterization of the functional role and biology of TAARs and identification of their endogenous and exogenous ligands will eventually promote these receptors as an attractive class of targets to correct monoaminergic processes that could be dysfunctional in a host of disorders of brain and periphery.

Cross-Talk between G Protein-Coupled Receptors: Challenges of Distinguishing Upstream from Downstream Mechanisms

Jonathan Javitch

College of Physicians and Surgeons, Columbia University, USA

A major obstacle to understanding the functional importance of dimerization between class A G protein-coupled receptors (GPCRs) has been the methodological limitation in achieving control of the identity of the components comprising the signaling unit. We have developed a functional complementation assay that enables such control, and we have used this approach to study human dopamine D2 receptor function. The minimal signaling unit, two receptors and a single G protein, is maximally activated by agonist binding to a single protomer, which suggests an asymmetrical activated dimer. Inverse agonist binding to the second protomer enhances signaling, whereas agonist binding to the second protomer blunts signaling. Ligand-independent constitutive activation of the second protomer also inhibits signaling. Thus, GPCR dimer function can be modulated by the activity state of the second protomer, which for a heterodimer may be altered in pathological states. Our new methodology also makes possible the characterization of signaling from a defined heterodimer unit. We have also developed a new method that combines protein complementation with resonance energy transfer to study conformational changes in response to activation of a defined G protein-coupled receptor heteromer, and we have applied the approach to the putative dopamine D1-D2 receptor heteromer. Remarkably, the potency of the D2 dopamine receptor (D2R) agonist R-(-)-10,11-dihydroxy-N-n-propylnoraporphine (NPA) to change the Gα(i) conformation via the D2R protomer in the D1-D2 heteromer was enhanced ten-fold relative to its potency in the D2R homomer. In contrast, the potencies of the D2R agonists dopamine and quinpirole were the same in the homomer and heteromer. Thus, we have uncovered a molecular mechanism for functional selectivity in which a drug acts differently at a GPCR protomer depending on the identity of the second protomer participating in the formation of the signaling unit--opening the door to enhancing pharmacological specificity by targeting differences between homomeric and heteromeric signaling. The existence of GPCR dimers and/or oligomers has been demonstrated in heterologous systems using a variety of biochemical and biophysical assays, and these interactions are the subject of intense research because of their potential role in modulating signaling and altering pharmacology. Nonetheless, evidence for the existence of receptor interactions in vivo is still elusive because of a lack of appropriate methods to detect them. We have adapted and optimized a proximity ligation assay (PLA) for the detection in brain slices of molecular proximity of two antigens located on either the same or two different GPCRs. Using this approach, we were able to confirm the existence of dopamine D2 and adenosine A2A receptor complexes in the striatum of mice ex vivo.

Growth Hormone Secretagogue Receptor (GHSR1a, aka Ghrelin Receptor) is an Allosteric Modulator of Dopamine Signaling in the Brain via Formation of GHSR1a:DRD2 and GHSR1a:DRD1 Heteromers

Roy G. Smith, Celine Ullrich and Andras Kern
Department of Metabolism and Aging, Scripps Research Institute Florida, USA

A small molecule (MK-0677) designed to rejuvenate the growth hormone axis was used to clone an orphan GPCR that mediates MK-0677 action (Howard et al. Science 273:974, 1996). The receptor was designated GHSR1a and transcripts were identified in pancreas, anterior pituitary gland, hypothalamus, hippocampus and mid-brain. Subsequently, an endogenous GHSR1a agonist was identified in stomach extracts (Kojima et al. Nature 402:656, 1999). Curiously, despite localization of expression of GHSR1a in the brain, ghrelin is not made there and none is detectable other than trace amounts in the hypothalamus. Hence, we asked what the function of GHSR1a expression in the brain might be. By generating transgenic mice with IRES-tau-GFP knock-in at the ghsr locus we identified GHSR1a producing neurons (Jiang et al. Mol. Endocrinol. 20:1772, 2006). A combination of GFP fluorescence and immunocytochemistry identified subsets of neurons that co-expressed GHSR1a + dopamine receptor-2 (DRD2) and others expressing GHSR1a + dopamine receptor-1 (DRD1). In HEK293 cells, co-expression of GHSR1a with DRD2 resulted in non-canonical signal transduction in response to dopamine (DA) resulting in mobilization of $[Ca2 +]i$. Co-expression of GHSR1a + DRD1 also resulted in non-canonical signaling in response to DA and mobilization of $[Ca2 +]i$. In the case of DRD2, Ca2 + release is mediated via $G\alpha i$ dependent upon $G\beta\gamma$ subunits, but with DRD1, Ca2 + release is dependent upon $G\alpha q$. Using GHSR1a, DRD2 and DRD1 tagged with SNAP- and CLIP- and time resolved fluorescence resonance energy transfer (Tr-FRET) we showed that non-canonical DA signaling is a consequence of formation of GHSR1a:DRD2 and GHSR1a:DRD1 heteromers. The possibility of a mechanism involving receptor crosstalk caused by basal activity of GHSR1a was ruled out by experiments with GHSR1a point mutants and selective Gα-protein knockdown. We next demonstrated in primary neuronal cultures DRD2 agonist-induced mobilization of [Ca2 +]i in neurons that co-express GHSR1a + DRD2. Similarly, in organotypic cultures of brain sections a DRD1 agonist induces [Ca2 +]i mobilization in neurons co-expressing GHSR1a + DRD1. In brain slices from WT mouse we demonstrated the presence of GHSR1a:DRD2 heteromers in hypothalamic neurons by confocal FRET using a combination of a monoclonal DRD2 antibody and red fluorescent ghrelin. Brains from ghsr-/- and drd2-/- were used as negative controls. To test for a functional role of these heteromers we treated fasted WT, ghrelin-/- and ghsr-/- mice with a DRD2 agonist and measured food intake. The DRD2 agonist suppressed feeding behavior in the fasted WT and ghrelin-/- mice, but ghsr-/- mice were resistant to the anorexigenic effect of the DRD2 agonist. Furthermore, treatment of WT mice with the selective neutral GHSR1a antagonist (JMV2959) made them refractory to the anorexigenic effects of the DRD2 agonist. Hence, GHSR1a alone, and not ghrelin, plays an important role in regulation of DA signaling. In conclusion, GHSR1a:DRD2 and GHSR1a:DRD1 heteromers exist in neurons of native brain tissue resulting in allosteric modification of DA signaling. We show DA signaling through these heteromers is blocked by a GHSR1a antagonist. Hence, treatment with a GHSR1a antagonist provides a selective way of blocking DA signaling in neurons expressing the heteromers,without affecting signaling in neurons expressing DRD2 or DRD1 alone. These results show potential opportunities for developing more selective therapeutic agents for treating psychiatric disorders involving abnormal DA signaling.

The support of NIH grant R01 AG19230 to RGS is gratefully acknowledged.

Regulation of β2Adrenergic Receptor Trafficking and Signaling by Ubiquitination and Deubiquitination

Sudha Shenoy

Duke University Medical Center, USA

Agonist-stimulation of the β2adrenergic receptor (β2AR) leads to heterotrimeric G protein activation followed by an increase in cAMP synthesis by adenylyl cyclase and downstream signaling. This is counteracted by receptor phosphorylation by G protein coupled receptor kinase(s) and subsequent high affinity binding of the adaptor proteins β-arrestin1 and 2. In addition to this desensitization of G protein signaling, β-arrestins facilitate β2AR internalization via clathrin-coated vesicles and initiate ERK1/2 phosphorylation in a G protein-independent manner. Agonist-activation also leads to ubiquitination of both the β2AR and β-arrestin leading to distinct functional outcomes: receptor ubiquitination leads to its lysosomal degradation, whereas β-arrestin ubiquitination facilitates receptor interaction, clathrin binding and scaffolding of ERK1/2. Ubiquitination of the β2AR and β-arrestin2 are mediated by distinct enzymes or E3 ubiquitin ligases: Nedd4 ubiquitinates the β2AR and Mdm2 modifies β-arrestin2. On the other hand de-ubiquitination of both β2AR and β-arrestin2 is carried out by a pair of de-ubiquitinases (DUBs), namely, USP20 and USP33. De-ubiquitination of β-arrestin2 renders the dissociation of β-arrestin2-β2AR complexes and curtails β-arrestin-dependent ERK1/2 activation, whereas de-ubiquitination of the agonist-stimulated β2AR prevents its trafficking to lysosomal compartments and promotes recycling and resensitization of cAMP signaling. Accordingly, reversible ubiquitination functions as a molecular mechanism that integrates β2AR trafficking and signaling and could play a critical role in balancing physiological responsiveness.

Sorting NEXIN-25 Interacts with D1 and D2Dopamine Receptors to Regulate Receptor Expression and Signaling

David R. Sibley, R. Benjamin Free, Yoon Namkung and A. Lisa

Hazelwood Molecular Neuropharmacology Section, NINDS, National Institutes of Health, Bethesda, MD, USA

Our studies employing co-immunoprecipitation assays for dopamine receptors (DARs) coupled with mass spectrometry-based sequencing have identified sorting nexin-25 as a member of the DAR signalplex. Mammalian sorting nexins (SNXs) have been suggested to regulate intracellular trafficking, internalization, and endosomal recycling or sorting of membrane-bound cargo. Thus far, 27 SNXs have been identified in humans, all defined by the presence of the phox (PX) domain. SNX25 also contains an RGS (regulator of G-protein signaling) domain. The physiological role of SNX25 is unknown. Using RT-PCR we have found that SNX25 is expressed in multiple tissues including brain and kidney and that the endogenous isoform is longer than the one indicated in GenBank. The full-length SNX25 contains two putative transmembrane-spanning regions and con-focal microscopy shows that it is localized in distinct clusters at or near the plasma membrane. When SNX25 is over-expressed with DARs, the receptors show changes in expression patterns and appear localized to the SNX25 clusters. Over-expression of SNX25 perturbed both endocytosis and recycling of the D2DAR. Radioligand binding also show that the expression levels of both D1and D2DARs are increased with SNX25 over-expression. Decreasing the levels of endogenous

SNX25 using siRNA causes a subsequent decrease in DAR expression. These data suggest that SNX25 plays a role in DAR trafficking through intracellular membrane compartments and regulates both receptor expression and signaling. Supported by The NIH intramural program.

Trafficking of the D2 Dopamine Receptor Promotes Drug-Induced Changes in Neuronal and Behavioral Plasticity

Pia Tschische, Anuradha Madhavan, Jegath Athiligam, Laura Milan-Lobo, Dawn Thompson, Linda Wilbrecht and Jennifer L. Whistler

University of California, San Francisco, Ernest Gallo Clinic and Research Center, University of California, San Francisco, Department of Neurology. 5858 Horton St. Suite 200, Emeryville, CA 94608, USA

Dysregulation of dopaminergic signaling is associated with several different neuropsychiatric diseases including Parkinson's disease, depression, bipolar disorder, schizophrenia and addiction. Dopamine mediates its effects through its action at five distinct receptors, D1, D2, D3, D4 and D5, belonging to the G protein-coupled receptor (GPCR) superfamily. These receptors can be subdivided into two groups: the D1-like (D1 and D5), which are coupled to the stimulatory G proteins Gs and Golf, and the D2-like (D2, D3, D4), which are coupled to the inhibitory G proteins Gi/o. While the molecular mechanisms responsible for the alterations in dopamine signaling in the disease states mentioned above are largely unknown, dopamine receptors, in particular D2-like receptors, are significantly downregulated in untreated schizophrenic and bipolar patients, in patients with chronic depression and, in drug abusers. Indeed, studies using positron emission topography have consistently shown that nicotine, heroin, alcohol, methamphetamine and cocaine abuse are accompanied by a decrease in striatal D2 receptor (D2R) availability.Drug-induced loss of D2R is apparent across multiple species including rodents, primates, and humans. More broadly, the large degree of co-morbidities of depression, anxiety, schizophrenia, bipolar symptoms and drug abuse have led to the hypothesis that downregulation of D2Rs is contributing to many modalities of neuropsychiatric disease. However, the molecular mechanisms that mediate loss of D2R in vivo remain unclear. Dopamine receptor-mediated signaling is extensively regulated by numerous processes. One way is by endocytosis whereby receptors are removed from the cell surface after activation. We have found that, following endocytosis, distinct dopamine receptors are sorted differentially: the D1 receptors (D1Rs) are recycled, while the D2Rs, are degraded. We also identified a protein, GPCR-associated sorting protein 1 (GASP1), which is responsible for the targeting of the D2R for degradation after endocytosis. Because the D1Rs and D2Rs have opposing effects on signal transduction and neuronal excitability, selective loss of D2Rs would be expected to alter responsiveness to dopamine. We propose that the balance of D1R-Gs signaling versus D2R-Gi signaling in circuits that express both of these receptors subtypes is disrupted due to downregulation of D2R under conditions of high dopamine tone, such as prolonged drug use. We also hypothesized that downregulation of D2Rs by GASP1 was responsible for some of the changes in neuronal and behavioral plasticity associated with drug use. Here we will describe our recent findings, which demonstrate that preventing downregulation of D2Rs in response to repeated cocaine, either through disruption of GASP1 function or by using ligands at the D2R biased for signaling but not trafficking of the D2R, disrupts both neuronal and behavioral plasticity associated with repeated drug use.

D2 Receptors Control of Dopamine Levels

Emiliana Borrelli

University of California Irvine, USA

Dopamine (DA) is a major neuromodulator of the central nervous system. DA signaling is critical in very diverse physiological functions, which range from locomotion to cognition, from reward and motivation to hormone synthesis and release. This wide variety of functions explains why dysfunctions of the dopaminergic system leads to major neurological and psychiatric human disorders, such as Parkinson's disease and schizophrenia. DA signaling is also involved in the behavioral and cellular effects induced by drugs of abuse. Thereby a thorough understanding of the mechanisms underlying DA dependent warrant in depth studies for future design of specific therapeutic and pharmacological intervention of human disorders. Our research focuses on studying the molecular mechanisms responsible for the behavioral and cellular effects induced by DA in vivo. Our research focuses onunraveling the role of DA D2 receptors (D2R) in vivo. Importantly, knockout of D2R generates a motor Parkinson's-like phenotype, leads to pituitary tumors, and abolishes the rewarding and reinforcing properties of opiates, among several other impairments. The outstanding phenotypes of D2R mutants are likely ascribed to the wide distribution of these receptors in the brain and to the multiple roles that they serve in vivo, which are not common to the other DA receptors. For instance D2Rs are responsible for the regulation of DA synthesis and release from dopaminergic neurons; in addition, they also inhibit the release of heterogeneous neurotransmitters from cortical and striatal neurons and interneurons. Moreover, two isoforms of D2Rs are present in the brain, D2L (long) and D2S (short), both isoforms are generated from the same gene by a mechanism of alternative splicing. Recently, we have developed novel D2R mutants with which address site- and isoform-specific functions. We find that the selective loss of D2 autoreceptors in the substantia nigra compacta and ventral tegmental area unmasks prominent feedback mechanisms regulating DA release. Importantly, we identified that these inhibitory feedback loops are

D2R-mediated and, interestingly, they have a relatively larger impact in the dorsal striatum than in the nucleus accumbens, in agreement with differential regulation of DA release in these areas. These results further our knowledge on the DAergic system and might provide insights on the physiopathological mechanisms underlying human pathologies and addiction.

Integrated Approaches to Understand the Actions of GPCRs: The b-Arrestin-Dependent D2R Mediated Signaling Through Akt/GSK3

Marc G. Caron, Nikhil M. Urs, Josh C. Snyder and Sean M. Peterson

Departments of Cell Biology, Medicine and Neurobiology Duke University Medical Center,
Durham NC 27710, USA

In the brain dopamine (DA) is an important modulator of fast neurotransmission and is implicated in the behavioral control of locomotion, cognition, affect, and reward. Dysregulation in these systems has been implicated in the manifestation of several neurological conditions. These actions of DA are mediated through activation of D1 and D2-like G protein-coupled receptors (GPCR). Like other GPCRs, D2R display functional selectivity by engaging Akt/GSK3 signaling in a G protein-independent way through the ability of b-arrestin2 to scaffold an Akt/PP2A/GSK3 complex. These biochemical interactions are targets of antipsychotics and lithium (Beaulieu et al., 2008; O'Brien et al., 2011). To further validate the biochemical, cellular and behavioral functions of this signaling pathway we are combining several genetic approaches to either delete components of this signal transduction pathway in a cell specific fashion or to reconstitute deletion of D2R in medium spiny neurons (MSN) of the striatum with engineered D2R that can signal selectively through G protein- or b-arrestin2-dependent mechanisms. In order to selectively inactivate GSK3b in striatal MSNs we have crossed GSK3bflox mice with D1RCre or A2ARCre mice, which respectively inactivate GSK3b in D1R- and D2R- expressing MSNs of the striatal complex. These mice were then analyzed in behaviors commonly used to test antipsychotic efficacy or behaviors that are sensitive to lithium treatment. Inactivation of GSK3b in D2R but not D1R expressing MSNs, recapitulates the effects of antipsychotics on behavior in the animals. Thus, D2RGSK3b-/-but not D1RGSK3b-/-mice showed a reduced locomotor response to amphetamine and rearing response to apomorphine, lack of pre-pulse inhibition disruption by amphetamine or apomorphine, and increased spontaneous alternations in the Y-maze, reminiscent of the actions of antipsychotics. However, haloperidol induced catalepsy and place preference to amphetamine was unchanged in either the D2RGSK3β-/-or D1RGSK3β-/-mice compared to control mice. Interestingly, stabilization of b-catenin, a downstream target of GSK3β, in D2R expressing MSNs did not affect any of the behaviors tested. Moreover, D2RGSK3β-/-or D1RGSK3β-/-mice showed similar responses to littermate controls in the tail suspension and dark-light emergence test, behaviors, which we previously demonstrated to be b-arrestin2 and GSK3b dependent and sensitive to lithium treatment. To complement this approach we have also engineered D2R mutants that retain membrane expression and agonist dependency but are functionally selective for G protein- or b-arretin2-dependent signaling. In vivo viral reconstitution of these selective D2Rs into mice lacking D2R in MSNs should recapitulate the above properties and facilitate the identification of downstream targets of the functionally selective D2Rs. Taken together these results suggest that deletion of GSK3bbut not stabilization of b-catenin in D2R MSNs mimics antipsychotic action without affecting signaling pathways involved in catalepsy or mood stabilization suggesting that these behaviors might be mediated through other neuronal pathways. These genetic approaches coupled with the feasibility to identify functionally selective ligands for each pathway (Allen et al., 2011) may provide new avenues for more effective and selective therapies for modulation of GPCR signaling.

Hyperdopaminergic Modulation of Inhibitory Transmission is Dependent on GSK-3β Signaling-Mediated Trafficking of GABAA Receptors

Yan-Chun Li, Min-Juan Wang and Wen-Jun Gao

Dept of Neurobiology, Drexel University College of Medicine, Philadelphia, PA, USA

BACKGROUND: Cortical dopamine (DA) modulation of the gamma-amino butyric acid (GABA) system is closely associated with cognitive function and psychiatric disorders. We recently reported that the glycogen synthase kinase 3β (GSK-3β) pathway is required for hyperdopamine/D2 receptor-mediated inhibition ofNMDA receptors in the prefrontal cortex.

OBJECTIVES: Here we explore whether GSK-3β is also involved in dopaminergic modulation of GABAA receptor-mediated inhibitory transmission.

METHODS AND RESULTS: We confirmed that DA induces a dose-dependent, bidirectional regulatory effect on inhibitory postsynaptic currents (IPSCs) in prefrontal neurons. The modulatory effects of DA were differentially affected by co-application of GSK-3β inhibitors and different doses of DA. GSK-3β inhibitors completely blocked high-dose (20 μM) DA-induced depressive effects on IPSCs but exhibited limited effects on the facilitating regulation of IPSC in low-dose DA (200 nM). We also confirmed that surface expressions of GABAAreceptor β2/3 subunits were significantly decreased by DA applied in cultured prefrontal neurons and in vivo administration of DA reuptake inhibitor. These effects were blocked by prior administration of GSK-3β inhibitors. We explored DA-mediated regulation of GABAAreceptor trafficking and exhibited the participation of brefeldin A-inhibited GDP/GTP exchange factor 2 (BIG2) or dynamin-dependent trafficking of GABAAreceptors.

CONCLUSIONS: Together, these data suggest that DA may act through different signaling pathways to affect synaptic inhibition, depending on the concentration. The GSK-3β signaling pathway is involved in DA-induced decrease in BIG2-dependent insertion and an increase in the dynamin-dependent internalization of GABAAreceptors, which results in suppression of inhibitory synaptic transmission.

Simultaneous, Massively Parallel and Genome-Wide Interrogation of GPCR b-arrestin-ergic Signaling: Implications for GPCR Drug Discovery and Catecholamine Biology

B.L. Roth, M.F. Sassano and W.K. Kroeze

Department of Pharmacology, Division of Chemical Biology and Medicinal Chemistry, and the National Institute of Mental Health Psychoactive Drug Screening Program, University of North Carolina Medical School, Chapel Hill, NC 27599, USA

GPCR functional selectivity—the process whereby ligands specifically engage distinct signaling pathways—is now a well-established and validated pharmacological concept. Thus, as we and many others have demonstrated, GPCR ligands can be identified which differentially activate canonical G-protein signaling pathways,b-arrestin-mediated signaling and non-G protein/non-arrest-in-ergic signaling. With respect to catecholamine receptors, we have identified ligands which appear to possess extreme patterns of functional selectivity with bias for either Gi- or b-arrestin signaling pathways (Allen et al, PNAS 2011; Chen et al, in revision; MF Sassano et al, this meeting). These and similar sorts of findings by many other investigators have inspired us to develop a genome-wide platform suitable for interrogating the entire complement of druggable, non-olfactory

GPCRs in a simultaneous and massively parallel fashion in order to identify new small molecules which can target GPCR-b-arrestin signaling. To accomplish this task we synthesized a codon-optimized library encompassing essentially all the non-olfactory, druggable GPCRs (NOD-GPCR) whereby each synthetic cDNA was expressed in-frame with: (1) a plasma membrane targeting sequence; (2) an epitope tag to verify surface expression; (3) a TEV protease site; (4) an arrestin-targeting sequence and (5) a transcription factor to promote luciferase expression. The general approach was similar to that originally described by Barnea et al (PNAS 2007) albeit with key and non-trivial modifications. Each construct was then individually transfected via a robotic liquid-handling system into each well of a 384-well plate. In this way, we can screen a single ligand simultaneously at the entire complement of NOD-GPCRs in two 384-well plates (including appropriate controls). We found that the vast majority of NOD-GPCRs, including ~160 orphan GPCRs, were surface-expressed and functional. I will present new and unpublished data whereby this unique resource has been used to profile large numbers of catecholaminergic ligands and will demonstrate that the overwhelming majority possess significant off-target b-arrestin signaling potentialities at the NOD-GPCR-ome. The relevance of these findings for catecholamine researchers and drug discovery will be highlighted. Supported by U19MH82441 and the NIMH PDSP.

Backwards and Forwards with Molecular and Classical Pharmacology

Brian Shoichet

Department of Pharmaceutical Chemistry, UCSF, USA

The recent determination of the structures of G-Protein Coupled Receptors enables the structure-based prediction of new ligands for these classic targets. Unbiased docking screens of over 3 Million ZINC library molecules against the ß2-Adrenergic, the Adenosine A2a, the Dopamine D3, and the Chemokine CXCR4 receptors, have returned new scaffolds and sub-micromolar affinity ligands with hit rates ranging from 17 to 35%. The structural, computational and chemical space origins of this hit rate will be discussed, as will opportunities for using homology models of these targets and for finding agonists as well as antagonists. In contrast to this molecular biology view of pharmacology, which begins with targets and seeks ligands, the pharmacology practiced in the classical period (~1930 to ~1985) began with ligands and used these to define receptors. As peculiar as this direction of information seems to us, in its day it was highly successful. Here we return tothis classic idea, seeking unexpected similarities. Using a chemoinformatic method, we quantitatively compare over 3000 targets to one another based on the similarity of their ligands. The relationships that emerge predict previously unknown off-targets for specific drugs and reagents. Applications to predicting adverse drug reactions, identifying the molecular targets in phenotypic screens, and to understanding the targets underlying mechanism of action will be considered.

Targeting Beta-Arrestin Dependent Signaling in the Treatment of Parkinson's Disease

Nikhil M. Urs[1], Tanya L. Daigle[1], Jonathan Ting[3] and Marc G. Caron[1,2,3]

[1]Departments of Cell Biology; [2]Medicine; [3]Neurobiology, Duke University Durham NC 27710, USA

BACKGROUND: Dopamine (DA) is a major catecholamine neurotransmitter that is released by midbrain DA neurons and activates G protein coupled receptors (GPCRs) on postsynaptic neurons. Recent studies from our laboratory with the two main classes of DA receptors (D1 and D2) have shown that in addition to canonical G protein signaling, some of their physiological functions such as locomotion are mediated through beta-arrestin2 (Barr2)-dependent mechanisms. In Parkinson's disease (PD) the severe locomotor deficits that include akinesia, bradykinesia and tremor are commonly treated with L-DOPA and/or dopamine receptor agonists. Although these treatments ameliorate the locomotor deficits, prolonged L-DOPA usage causes L-DOPA induced dyskinesias (LIDs). Recent studies in animal models of PD have suggested that dyskinesias might be associated with G protein-mediated signaling at D1 and D2 receptors. Therefore, preventing G protein signaling butenhancing beta-arrestin dependent DA receptor signaling through pharmacological intervention may be a novel way to elicit desirable DA-dependent behaviors such as locomotion that are beneficial to Parkinsonian patients without eliciting dyskinesias.

OBJECTIVES: Our objective is to test the effect of over-expression or deletion of Barr2 on L-DOPA induced forward locomotion and dyskinesia potential. We also want to determine the DA receptor subtypes that are involved in manifestation of forward locomotion and dyskinesias. We hypothesize that over-expression of Barr2 in striatal neurons of mice will enhance the efficacy of L-DOPA mediated locomotion while reducing dyskinesias whereas deletion of Barr2 will impair L-DOPA mediated locomotion but increase the occurrence of dyskinesias.

METHODS AND RESULTS: In this study we have utilized an acute DA-deficient DAT-KO (DDD) mouse model that mimics PD symptoms. To over-express Barr2 we have utilized a viral approach whereas to delete Barr2 we crossed the DAT-KO mice to the Barr2-KO mice to generate the DDD/WT and DDD/Barr2KO mice. We have obtained preliminary evidence that suggests that beta-arrestin2 (Barr2) is required for L-DOPA induced locomotion.Acute L-DOPA (25 mg/kg, i.p) injections induced forward locomotion in the DDD/WT mice but interestingly the DDD/Barr2KO mice show reduced forward locomotion and enhanced three-paw dyskinesia activity. In the DDD/WT mice, chronic treatment with L-DOPA (9 injections) gradually increased dyskinesias but inhibited forward locomotion. Interestingly, acute injection of extremely high doses of L-DOPA (100-200 mg/kg) also induced three-paw dyskinesias in mice. Conversely, over-expression of Barr2 in the striatum of the DDD/WT mice increased the efficacy of forward locomotion upon chronic exposure to L-DOPA. Administration of D1/D5 but not D2/D3 receptor antagonists inhibited dyskinesias and forward locomotion. To determine the role of D5 receptors in L-DOPA induced behaviors we crossed the DAT-KO to the D5-KO mice to generate the DDD/D5WT and DDD/D5KO mice. Interestingly, upon acute L-DOPA injection the DDD/D5WT mice showed forward locomotion whereas the DDD/D5KO mice showed enhanced dyskinesias.

INTERPRETATIONS AND CONCLUSIONS: These data suggest that while the absence of Barr2 reduces the beneficial locomotor response of L-DOPA, it enhances the manifestation of a L-DOPA induced behavior akin to dyskinesias, presumably because it shifts the system to excessive G protein-dependent signaling. Interestingly, over-expression of Barr2 enhances the efficacy of forward locomotion. Moreover, inhibition of D5 receptor signaling enhances acute L-DOPA induced dyskinesia that mimics the DDD/Barr2KO mice. Together these results support a role for Barr2 and D5 receptors in L-DOPA induced forward locomotion but inhibition of dyskinesias.

D2 Functionally Selective Ligands: novel Therapeutics?

Maria F. Sassano[1]**, John A. Allen**[2]**, Jian Jin**[3]**, Vincent Setola**[1]**, William C. Wetsel**[4]**,
Marc G. Caron**[5] **and Bryan L. Roth**[1]

[1]*Departments of Pharmacology and National Institute of Mental Health Psychoactive Drug Screening Program,
School of Medicine, University of North Carolina, Chapel Hill NC, 27599;* [2]*Pfizer Inc, Neuroscience Medicinal
Chemistry, Groton, CT 06340;* [3]*Center for Integrative Chemical Biology and Drug Discovery, Division of Chemical
Biology and Medicinal Chemistry, Eshelman School of Pharmacy, University of North Carolina,
Chapel Hill NC, 27599;* [4]*Departments of Psychiatry and Behavioral Sciences, Cell Biology, and Neurobiology*
[5]*Departments of Cell Biology, Medicine, and Neurobiology, Duke University Medical Center, Durham, NC 27710.*

β-arrestin−biased GPCR ligands that selectively activate β-arrestin signaling pathways over Gi-coupled pathways have not been reported in the past. Aripiprazole, an FDA-approved atypical antipsychotic drug, was one of the first functionally selective D2 receptor (D2R or D2) ligands identified (Mailman and Murthy; Lawler, Prioleau et al. 1999; Urban, Vargas et al. 2007).

Understanding which signaling pathways contribute selectively to antipsychotic efficacy and side effects will allow for the design of better antipsychotic drug candidates and, ultimately, lead to safer and more effective therapies for patients. Structure−functional−selectivity relationships (SFSR) of the aripiprazole scaffold have not been studied and only modest structure activity relationships (SAR) have been reported (Johnson, Choi et al.; Oshiro, Sato et al. 1998). Our first findings show several β-arrest-in−biased GPCR ligands that selectively activate β-arrestin signaling pathways over Gi-coupled. Ongoing efforts are focused on finding new functionally selective aripiprazole-like ligands, especially, or most interest, to find ligands that are functionally selective for Gi-coupled over β-arrestin for the D2 receptor.

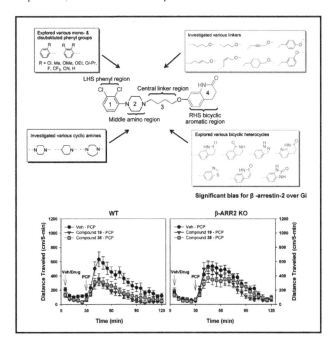

To test compounds for functional selectivity, we used well-established cell-based functional assays that measure D2-Mediated Gi-Regulated Inhibition of cAMP Production (GloSensor assay) and D2-Mediated β-Arrestin-2 Translocation (Tango assay). Orthologous assays (D2-mediated β-arrestin-2 translocation by enzyme fragment complementation and D2-mediated BRET-based β-arrestin-2 recruitment) were performed to validate these findings. To test efficacy of β-arrest-in-biased compounds, locomotor activity was assessed in wild-type (WT) or β-arrestin-2 knockout (β-ARR2 KO). We demonstrate, using wild type and β-arrestin-2 knockout mice, that the atypical antipsychotic- like profile of these β-arrestin−biased D2R agonists requires β-arrestin-2.

Very little attention has been devoted to creating and annotating ligands with distinct patterns of functional selectivity. The development of these unique, func-tionally selective, β-arrestin−biased and Gi-biased D2R ligands provide the biomedical community with valuable chemical tools for probing signaling pathways essential for antipsychotic efficacy and side effects.

Investigation of the D_1-D_2 heteromer: pharmacology, subunit composition, and mechanisms of signaling.

Lani S. Chun[1,2], R. Benjamin Free[1], Trevor B. Doyle[1], Xi-Ping Huang[3] and David R. Sibley[1]

[1]NINDS-IRP, Bethesda, Maryland, United States; [2]CMDB Program, Johns Hopkins University, Baltimore, Maryland, United States; [3]Dept. of Pharmacology, UNC School of Medicine, Chapel Hill, NC, United States

BACKGROUND: The D_1 (D_1R) and D_2 (D_2R) dopamine (DA) receptors (DARs) modulate cAMP levels via activation of G_s and G_i proteins. D_1Rs and D_2Rs are also thought to form hetero-oligomers. Studies have found that, when co-activated, the heteromer couples to non-canonical signaling pathways involving phospholipase C (PLC) activation and Ca^{2+} mobilization. SKF83959, a substituted benzazepine, has even been suggested to be a D_1-D_2 heteromer-selective agonist and a functional probe of the heteromer *in vivo*.

OBJECTIVES: We investigated several pharmacological and mechanistic aspects of the D_1-D_2 heteromer model.

METHODS AND RESULTS: The $G_{\alpha q}$ subunit, D_1R, D_{2L}R, and D_{2S}R were transiently transfected into HEK293T cells. Ca^{2+} mobilization was measured kinetically using Fluo-8 and a Hamamatsu FDSS/μCell plate reader. DA elicited a robust Ca^{2+} response in cells co-transfected with the D_1R + D_{2S}R or D_1R + D_{2L}R. No Ca^{2+} response was observed in cells singly expressing either receptor. SKF83959 did not induce a Ca^{2+} response in any transfection condition and was found to exhibit sub-micromolar affinity for a variety of monoamine targets including some serotonergic, adrenergic, and dopaminergic receptors, as well as the serotonin transporter. A Ca^{2+} response to DA was elicited in D_1R + $G_{\alpha q}$ co-transfected cells and boosted in D_1R + D_{2L}R + $G_{\alpha q}$ co-transfected cells. The Ca^{2+} response could be drastically inhibited by pretreating the cells with $G_{i/o}$ or G_s protein toxins (pertussis toxin or cholera toxin, respectively), or by co-transfecting the DARs with catalytically inactive GRK2 mutants which sequester $G_{\beta\gamma}$ subunits.

CONCLUSIONS: Our results show that D_1R co-expressed with either D_2R isoform can promote Ca^{2+} mobilization. In contrast to previously published results, SKF83959 is not a D_1-D_2 heteromer agonist and does not appear to be useful as and *in vivo* probe. Finally, the mechanisms of D_1R + D_2R-mediated Ca^{2+} mobilization appear to be significantly more complicated than a linear heteromer − $G_{\alpha q}$ − PLC activation pathway. *Supported by The NIH intramural program.*

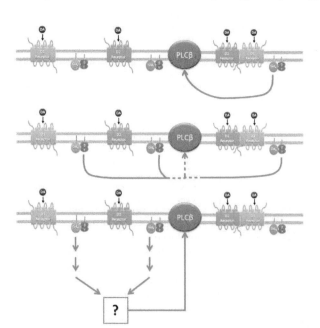

β-Adrenergic Receptors, Caveolae and Caveolins: Receptor Localization and Function

Paul A. Insel[1,2], **Stephen R. Fuhs**[1], **Anna Busija**[1,3], **David M. Roth**[3] **and Hemal H. Patel**[3]

[1]*Departments of Pharmacology;* [2]*Medicine;* [3]*Anesthesiology, UCSD, La Jolla, CA, 92093, USA*

Abundant data in recent years have demonstrated that: 1) numerous G-protein-coupled receptors (GPCRs), including β1- and β2-adrenergic receptors (ARs), localize in plasma membrane lipid raft microdomains that are enriched in cholesterol and certain other lipids and 2) caveolae, which are raft domains that also contain the protein caveolin (cav, of which there are 3 isoforms, cav-1, cav-2 and cav-3) contribute to signal transduction and agonist-mediated regulation of GPCRs. However, the precise molecular mechanisms that determine why only certain GPCRs interact with cavs are poorly defined; moreover, β1ARs and β2ARs show differences in their localization in caveolae. The functional roles of these receptor-caveolin interactions *in vivo* are also not well understood. We have taken a multi-pronged approach to address these issues. We discovered that caveolins can be SUMOylated and identified the preferred lysine residue (Lys[38]) on cav-3 that undergoes SUMOylation; we also defined the enzymes involved in SUMOylation of cav-3. Importantly, we found that the absence of SUMOylation alters the expression and stability of β2AR but not β1AR, apparently as a consequence of altered agonist-promoted desensitization. In other studies, we overexpressed cav-3 (a cav that shows muscle-specific expression) in cardiac myocytes of mouse hearts and found that the hearts from these mice have greater contractile and relaxation response to isoproterenol, a decrease in age-related loss in response to isoproterenol and greater exercise capacity. Taken together, our data indicate that both the cellular level of cav-3 expression and its ability to undergo SUMOylation contribute to the functional activity of βARs, thus influencing cellular and tissue response to catecholamines.

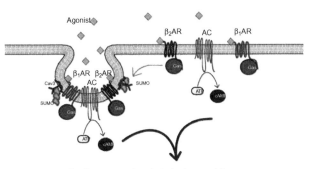

Cardiac response to catecholamine agonists

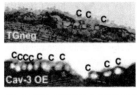

EM of cardiac membranes in TGneg and cardiac-specific Cav-3 overexpressing (Cav-3 OE) mice.

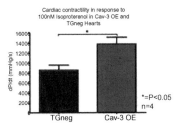

Cardiac contractility in response to 100nM Isoproterenol in Cav-3 OE and TGneg Hearts

$*=P<0.05$
n=4

(Supported by research and training grants from NIH).

Protein Kinase C β and the Dopamine Transporter Influence Surface D2-Like Dopamine Receptor Localization

Kathryn Luderman[1], Rong Chen[1], Paul Alberts[2] and Margaret Gnegy[1]

[1]University of Michigan, Ann Arbor, Michigan, United States; [2]University of Ottawa, Ottawa, Ontario, Canada

BACKGROUND: The dopamine transporter (DAT) and D2-Like dopamine autoreceptor (D2DR) regulate extracellular dopamine levels, and thus dopamine signaling. D2DR and DAT couple between the third intracellular loop of D2DR and the N-terminus of DAT. D2DR regulation of DAT is well known but, regulation of D2DR by DAT is less understood. Both D2DR and DAT contain protein kinase C phosphorylation sites. Protein Kinase C β (PKCβ) phosphorylates both D2DR and DAT to regulate the activity and/or trafficking of each molecule.

OBJECTIVE: Our objective was to determine if DAT and PKCβ regulate surface D2DR localization and if the regulation occurs through phosphorylation of DAT or D2DR.

METHODS: Neuroblastoma N2a cells were transfected with N-terminal FLAG-tagged D2DR and extracellular loop 2-HA-tagged DAT or only FLAG-D2DR with vector control. Basal and D2DR agonist (quinpirole)-stimulated surface localization for DAT and D2DR were determined. Surface D2DR was measured using immunofluorescence labeling using antibodies against the tagged proteins. PKCβeffects on D2DR surface localization were determined using the specific inhibitor LY379196 and mutant D2DR which lacks three putative PKC phosphorylation sites in the third intracellular loop. D2DR-DAT coupling was also assessed using a truncation mutant of DAT.

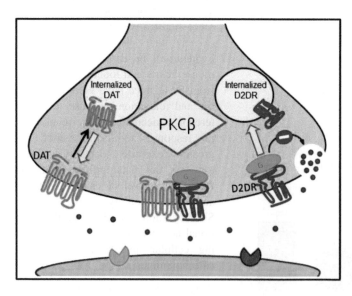

RESULTS AND CONCLUSIONS: Treatment with quinpirole and LY379196 both significantly change surface D2DR localization. These results differ depending on DAT presence in the cell. The PKCβ inhibition effect on D2DR surface localization was blocked by removing putative PKC phosphorylation sites as well as using a DAT truncation mutant. This study suggests that DAT and PKCβ influences D2DR surface localization, which may occur through several mechanisms that are the source of ongoing work. This work has broad implications as D2DR and DAT control the level of extracellular dopamine, which is an important factor in drug addiction and other mental diseases.
Supported by NIH grants DA 011697, DA 007267, and GM 07767.

Role of Adrenaline on the Maturation of β_2-Adrenoceptors

M. Moreira-Rodrigues[1,2], V. Monteiro[2], M. Morato[2,3], S. N. Ebert[4] and D. Moura[1,2]

[1]*Neuropharmacology, IBMC;* [2]*Department of Pharmacology and Therapeutics, FMUP;* [3]*Laboratory of Pharmacology, FFUP and REQUIMTE, Porto, Portugal;* [4]*Burnett School of Biomedical Sciences, UCF College of Medicine, Orlando, USA.*

BACKGROUND: It was suggested that there is a link between adrenaline and the maturation of β_2-adrenoceptor-mediated effects.

OBJECTIVES: The aim of this study was to characterize the role of adrenaline on the maturation of postjunctional β-adrenoceptor-mediated effects (smooth muscle relaxation) in PNMT (phenylethanolamine-N-methyltransferase, enzyme that transforms noradrenaline in adrenaline)-knockout mice.

METHOD AND RESULTS: Aortas of anesthetized (isoflurane) PNMT-knockout mice (8 to 12 weeks-old) were dissected, collected and cut into rings, which were mounted in a myograph. We evaluated dose-response curves to terbutaline (β_2-adrenoceptor agonist) in the absence or presence of ICI118551 (β_2-adrenoceptor antagonist), and to dobutamine (β_1-adrenoceptor agonist), in rings pre-contracted with phenylephrine. Adrenal glands were collected and immersed in perchloric acid (0.2 M). Adrenaline and noradrenaline were separated by reverse-phase HPLC and quantified by electrochemical detection. We confirmed that adrenaline is absent in the adrenal glands of PNMT-knockout mice. The potency and maximal effect of terbutaline were higher in controls than in PNMT-knockout mice ($EC_{50} = 5.2 \pm 2.8$ vs 29.0 ± 6.1 μM and $E_{max} = 17.7 \pm 3.3$ vs 7.0 ± 2.4 mN/mg, respectively). ICI11855 antagonized the effect of terbutaline in both groups, but the pA_2 was lower in PNMT-knockout than in control mice. The potency and the maximal effect of dobutamine were identical among groups ($EC_{50} = 0.22 \pm 0.07$ vs 0.18 ± 0.05 μM and $E_{max} = 20.0 \pm 4.2$ vs 13.8 ± 3.0 mN/mg, respectively).

CONCLUSIONS: In conclusion, when adrenaline is absent there is a lack of aortic postjunctional β_2-adrenoceptor-mediated vasodilation (but not of postjunctional β_1-adrenoceptor-mediated effects), which confirms the link between adrenaline and the maturation of β_2-adrenoceptor-mediated effects.

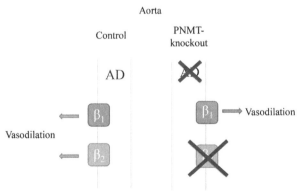

AD- Adrenaline
PNMT- phenylethanolamine-N-methyltransferase

Deciphering the Functionally Selective Properties of D2R Ligands

Prashant C. Donthamsetti and Jonathan A. Javitch

Columbia University, New York, NY, United States

BACKGROUND: Aripiprazole, the prototypical third generation antipsychotic, is marketed as a Dopamine Receptor 2 (D2R) partial agonist that stabilizes dopaminergic transmission in the mesolimbic and mesocortical systems of the brain, thus serving as a model for the development of D2R partial agonists for treatment of schizophrenia. However, in light of inconsistencies with this proposed model and the failure of other partial agonists in clinical trials, aripiprazole's molecular mechanism of action requires reevaluation. Recent evidence indicates that aripiprazole and other D2R ligands are functionally selective in activating pre- and postsynaptic D2R-mediated pathways of the mesolimbic system. Although functional selectivity has been predominantly attributed to biased signaling for either G-proteins versus arrestins, the molecular basis for the actions of aripiprazole and other D2R ligands have yet to be determined.

OBJECTIVES: To determine the molecular basis of functional selectivity for reported functionally selective D2R ligands by pharmacologically dissecting ligand-specific action at all relevant D2R coupling partners, including G-protein and arrestin subtypes.

METHODS AND RESULTS: In addition to utilizing an established bioluminescence resonance energy transfer (BRET)-based arrestin assay, a BRET-based system was developed capable of monitoring subtype specific G-protein coupling to D2R. Subtype specificity was achieved by tagging $G\alpha$ subtypes with a donor chromophore (Rluc8) and $G\beta$ and $G\gamma$ subtypes with a complemented split acceptor chromophore (Venus). Within this system, agonist-induced activation of D2R leads to dissociation of the heterotrimeric G-protein and thus a detectable change in the proximity of the sensors. The use of the split acceptor allows us for the first time to study signaling by a G-protein heterotrimer consisting of defined $G\alpha$, $G\beta$, and $G\gamma$, filtering out any contribution from endogenous G-proteins that might complicate the analysis.

CONCLUSIONS: To address the molecular basis of functional selectivity of D2R ligands, the G-protein BRET system was first validated pharmacologically. This system behaves as a typical functional assay in that stimulus- response coupling is sensitive to receptor and transducer level. In addition to subtype specificity, this system affords benefits over the canonical G-protein coupling assay, [35S] GTP-γS incorporation, in that signal output can be dynamically modulated over long time courses. A variety of functional G-protein sensors have been developed, including $G\alpha i/o/z$, $G\beta 1/2/5$, and $G\gamma 1/2/7/11$. Dose response curves have been generated for panel of 12 ligands, including aripiprazole, for multiple distinct heterotrimeric G-protein combinations. Whereas there is no significant difference in ligand activity by varying the $G\gamma$ subtype, partial agonists tend to have significantly higher efficacy in activating heterotrimers containing $G\alpha o$ and $G\alpha z$ compared to those with $G\alpha i$. Aripiprazole and other reported functionally selective D2R ligands are partial or full agonists at arrestin. To determine if these findings provide the mechanistic basis for functional selectivity seen at D2R-mediated mesolimbic pathways, the signaling roles of G-protein and arrestin subtypes must also be determined in native settings.

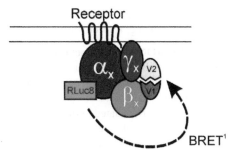

Chronic loss of Noradrenergic Tone Produces β-Arrestin2-Mediated Cocaine Hypersensitivity and a Gα$_I$ to Gα$_S$ switch in D2 Receptor Coupling in the Nucleus Accumbens

Meriem Gaval-Cruz[1], Brandon Goertz[2], Daniel J. Puttick[1], Rebecca C. Meyer[4], Dawn E. Bowles[3], Randy A. Hall[4], Carlos A. Paladini[2] and David Weinshenker[1]

[1]*Department of Human Genetics, Emory University, Atlanta, GA;* [2]*Department of Biology, UTSA, San Antonio, TX;* [3]*Department of Surgery, Duke University, Durham, NC;* [4]*Department of Pharmacology, Emory University, Atlanta, GA*

BACKGROUND: Chronic genetic or pharmacological inhibition of dopamine β–hydroxylase (DBH), the enzyme that converts dopamine (DA) to norepinephrine (NE) in noradrenergic neurons, enhances behavioral responses to cocaine and to a D2, but not D1, DA receptor agonist.

OBJECTIVES: To identify the cellular and molecular mechanisms underlying this behavioral hypersensitivity to cocaine and D2 activation.

METHOD AND RESULTS: We used western blotting to compare the relative abundance of dopamine receptor signaling proteins in the nucleus accumbens (NAc), caudate putamen, and prefrontal cortex of mice with chronic genetic or pharmacological DBH inhibition. We found that control mice treated chronically with the selective DBH inhibitor, nepicastat, and DBH knockout (*Dbh -/-*) mice have decreased β-arrestin2 and increased ΔFosB levels in the NAc, while all other D2 signaling proteins were unaltered. Viral vector-mediated overexpression of β-arrestin2 in the NAc normalized cocaine-induced locomotion in *Dbh -/-* mice. To investigate the cellular mechanisms underlying the D2 and cocaine hypersensitivity, we performed slice electrophysiology from NAc medium spiny neurons (MSN). As expected and reported previously, activation of D2R by quinpirole reduced excitability in MSN following current injection in slices from control mice. Remarkably, quinpirole increased excitability in MSN from *Dbh -/-* slices, an effect that was abolished by the selective Gαs inhibitor NF449. NF449 had no effect on quinpirole-induced inhibition in control mice. The Gαi inhibitor NF023 shifts the quinpirole response from inhibition to activation in neurons from control mice, but has no effect in those from *Dbh -/-* mice.

INTERPRETATION AND CONCLUSIONS: These results indicate that a decrease of β-arrestin2 in the NAc is necessary for cocaine hypersensitivity following chronic DBH inhibition, and identify a novel Gαi-to-Gαs switch in D2R coupling in MSN. In normal conditions, NE provides excitatory drive onto midbrain DA neurons, β-arrestin2 levels in the NAc MSN are normal, and D2R are coupled to Gαi/o. D2R stimulation inhibits their firing. In our proposed model, chronic DBH inhibition reduces NE drive onto midbrain DA neurons, decreasing their activity. In turn, this decreases β-arrestin2 abundance and D2 receptors become coupled to Gαs. Stimulation of D2 receptors now facilitates the firing of MSN in the NAc. The findings have implications for drug addiction, given the clinical efficacy of DBH inhibitors in the treatment of cocaine dependence.

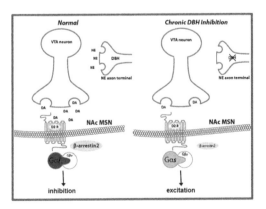

Probing the Functional Selectivity of β-adrenergic Receptors Reveals New Signaling Modes and Potential Therapeutic Applications

Emma van der Westhuizen[1], **Wayne Stallaert**[1], **Brigitte Murat**[1], **Martin Audet**[1], **Segolène Galandrin**[1], **Monique Lagacé**[1] and **Michel Bouvier**[1]

[1]*Department of Biochemistry and Institute for Research in Immunology and Cancer, Université de Montréal, Canada*

BACKGROUND: In recent years, it has become clear that GPCRs are not uni-dimensional switches that turn 'on' or 'off' single signaling pathways. Instead, each receptor can engage multiple signaling cascades that may or may not involve G protein activation. Individual ligands can have differential efficacies toward specific subsets of these signaling effectors [1,2]. This phenomenon known as ligand-biased signaling or functional selectivity offers interesting opportunities to develop compound with increased selectivity profiles but present important challenges for the drug discovery process [3].

OBJECTIVES: Using fluorescence and bioluminescence resonance energy transfer (FRET-BRET)-based biosensors that we recently developed [4,5,6,7] as well as label-free impedance measurements [8], the present study was aimed at monitoring multiple signaling pathways triggered by the β-adrenergic receptors in living cells and to assess the molecular determinants of ligand-biased signaling.

METHODS AND RESULTS: Using BRET- and FRET-based assays that monitor multiple signaling pathways we unraveled new signaling modes for β-adrenergic receptors (βAR) ligands and classified them into distinct subclasses that display unique signaling efficacy profiles toward different proximal effectors (Gi, Gs, βarrestin, etc..) and downstream signaling events (cAMP, MAPK, Calcium). The compound with distinct signaling profiles formed clusters of ligands that can also be differentiated based on the cellular impedance responses that they promote. The functional selectivity of ligands acting through a unique GPCR can therefore be assessed using either pathway specific biosensors or global label-free methods thus offering different avenues for drug candidate profiling. Such ligand-biased signaling profiles were observed in both engineered and native cells such as smooth muscle cells and cardiomyocytes indicating that they reflect intrinsic properties of the ligands. Direct assessment of the signaling molecular complexes involved, combined with molecular modeling and virtual docking of the ligands started to reveal the structural determinants of functional selectivity. These studies should provide the basis for the rational design of drugs with predetermined biased signalling profiles and improved therapeutic activities.

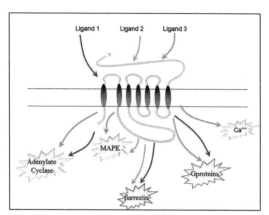

References

[1] Azzi M. *et al.* Proc. Natl. Acad. Sci. (USA), 100: 11406, 2003
[2] Galandrin S. *et al.* Trends Pharmacol. Sci. 28: 423, 2007
[3] Kenakin TP. Trends Pharmacol. Sci. 28:407, 2007
[4] Angers, S. *et al.* Proc. Natl. Acad. Sci. (USA), 97: 3684, 2000
[5] Charest, P.G. *et al.* EMBO Report, 6: 334, 2005
[6] Galés, C. *et al.* Nat. Struct. Mol. Biol., 13:778, 2006
[7] Breton, B. *et al.* Biophys. J., 99: 4037, 2010
[8] Stallaert, W. *et al.* PLoS One, 7:e29420, 2012

Identification of Substituted Benzazepines as Functionally Selective Ligands of the D_1 Dopamine Receptor

Jennie L. Conroy, R. Benjamin Free, Trevor B. Doyle and David R. Sibley

NINDS-IRP, Bethesda, Maryland, United States

BACKGROUND: The D_1 dopamine receptor (DAR) has been implicated in numerous neuropsychiatric disorders, and various D_1 ligands have shown great potential as therapeutic agents. However, numerous side effects have limited their clinical utility. Functionally selective compounds present a unique therapeutic opportunity to target individual pathways while minimizing signaling through others.

OBJECTIVE: A series of substituted "atypical" benzazepines, and D_1 "typical" agonists, were tested for their functional effects on D_1-mediated cAMP accumulation, D_1-mediated β-arrestin recruitment, and D_1 receptor internalization using live cell functional assays.

RESULTS: With respect to β-arrestin recruitment, the "atypical" agonists were found to be antagonists of this response whereas the "typical" agonists were nearly full agonists with efficacies ranging from 20-110%. We also examined D_1 receptor internalization using a novel β-galactosidase complementation assay. As with the β-arrestin recruitment assay, the "atypical" exhibited little agonist efficacy, yet the other "typical" were full agonists in terms of inducing receptor internalization. Our study also used two different D_1 cAMP assays: DiscoveRx (DRX) HitHunter, an antibody-based assay that provides a direct measurement of cAMP accumulation, and Codex, which utilizes a cyclic-nucleotide gated channel to measure changes in membrane potential as an indirect readout of cAMP production. In both DRX HitHunter and the Codex assay, all compounds elicited an increase in D_1 cAMP accumulation similar to that produced by dopamine, with efficacies ranging from 90-110% for the "typical" agonists and 50-100% for the "atypical" agonists, effects that were inhibited by the D_1 antagonist SCH23390.

CONCLUSIONS: Taken together, these data identify a novel group of substituted benzazepines that are functionally selective for the cAMP-mediated signaling pathway of the D_1 receptor. These data will be useful in interpretation of many seemingly contrasting in vivo/in vitro effects of these compounds, and may be useful in further identification of pathway-selective ligands of the D_1 dopamine receptor.

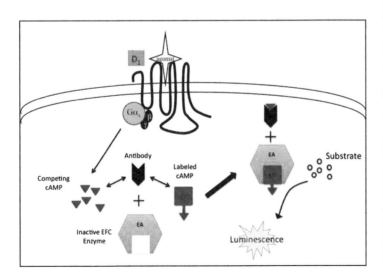

Novel cAMP Sensor Links GPCR-Gs Signaling to ERK in Neuroscreen-1 Cells

Andrew C. Emery[1] **and Lee E. Eiden**[1]

[1]*NIMH-IRP, Bethesda, Maryland, United States*

BACKGROUND: We have investigated the intracellular signaling pathways downstream of PACAP-mediated activation of the PAC1 receptor in neuroendocrine cells, where PACAP exerts its stress-transducing effects. PACAP causes pleotropic physiological effects, many of which require cAMP elevation. Canonically, cAMP signaling is PKA-dependent, however noncanonical (i.e. PKA-independent) cAMP pathways are an emerging concept in signaling.

OBJECTIVES: These studies pharmacologically define the cAMP-dependent signaling pathways PACAP employs to exert its effects in neuroendocrine cells.

METHODS AND RESULTS: PACAP acts through PAC1 receptors to stimulate cAMP elevation (EC50 = 17.31 + 3.31 nM), ERK phosphorylation, and neuritogenesis in Neuroscreen-1 (NS-1) cells. Cyclic AMP-elevating agents cholera toxin (50 µg/ml), forskolin (25 µM), and cAMP mimicking agents (db-cAMP, 8-Br-cAMP, and 8-CPT-cAMP; 500 µM) also stimulate ERK phosphorylation and neuritogenesis. Neuritogenesis and ERK phosphorylation induced by either PACAP or forskolin was blocked by the adenylate cyclase (AC) inhibitor 2'5'-dideoxyadenosine (100 µM). Neuritogenesis caused by all agents was blocked by U0126 (10 µM), implicating MEK/ERK in this process. While ERK phosphorylation and neuritogenesis due to treatment with PACAP or forskolin are clearly AC/cAMP-dependent, they are both PKA-independent processes, insensitive to PKA inhibitors H-89 (30 µM), KT5720 (30 µM), and PKI (100 nM). In contrast, PACAP and forskolin-induced CREB reporter gene transactivation and phosphorylation at serine 133 are potently blocked by PKA inhibitors. The Epac-selective cAMP analog 8-pCPT-2'-O-Me-cAMP supported Rap1A/B activation, but failed to support ERK phosphorylation and neuritogenesis.

CONCLUSIONS: PACAP, acting through the secretin-family GPCR PAC1, stimulates at least two functionally insulated cAMP-dependent signaling pathways: PKA-independent ERK phosphorylation and neuritogenesis, and PKA-dependent CREB phosphorylation, in NS-1 cells. Accessibility of these divergent cAMP-dependent signaling pathways to other GPCRs linked to cAMP elevation via Gs, including catecholamine (family A) receptors, is under investigation.

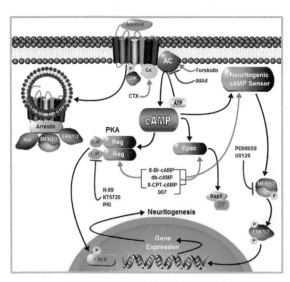

300-word text limit; graphical abstract (jpeg, tiff or eps)

~30% total space; 12-point.

High-Throughput Screening for Modulators of the D$_2$ Dopamine Receptor Yields Unique and Selective Pharmacological Chemotypes

R. Benjamin Free[1], Jennie L. Conroy[1], Rebecca A. Roof[1], Trevor Doyle[1], Yang Han[3],
Prashant Donthamsetti[3], Jonathan A. Javitch[3], Noel Southall[2], Kyle Emmitte[4], Craig Lindsley[4],
Marc Ferrer[2] and David R. Sibley[1]

[1]*NINDS/National Institutes of Health, Bethesda, MD;* [2]*NCATS/National Institutes of Health, Bethesda, MD;*
[3]*Department of Psychiatry/Columbia University, New York, NY;* [4]*Vanderbilt University Medical Center,*
Center for Neuroscience Drug Discovery, Nashville, TN

BACKGROUND: The D$_2$ dopamine receptor (DAR) is central in the etiology and/or therapy of many neuropsychiatric disorders. Specifically, D$_2$ DAR antagonism is the hallmark of all FDA-approved antipsychotics and stimulation of the D$_2$ DAR is critical for effective antiparkinsonian therapy. Unfortunately, truly specific drugs for this receptor have been difficult to obtain, primarily due to high conservation of the orthosteric binding site within DAR subtypes and among other G protein-coupled receptors.

OBJECTIVES: To interrogate a small molecule library for identification of novel small molecule scaffolds with desired functional characteristics, and define their mechanisms of action and selectivity among DAR subtypes.

METHODS: We developed a high throughput-screening (HTS) platform to interrogate large chemical compound libraries and screened a 370,000 + small molecule library to identify agonists, positive allosteric modulators, or antagonists. The primary HTS assay utilizes a cell line expressing the D$_2$ DAR coupled to a chimeric Gqi5 protein, thereby linking receptor activation to robust Ca^{2+} mobilization that is measured using a fluorescent readout (**see graphic**).

RESULTS AND DISCUSSIONS: The primary HTS-screen resulted in the identification of ~2,288 compounds with agonist activity and ~2,294 compounds with antagonist activity. Hits were subjected to orthogonal and counter-screening functional assays. While the primary goal for this screen was to identify allosteric compounds, a by-product of this screen and subsequent triaging was the identification of compounds that, while orthosteric, exhibit high selectivity for the D$_2$ DAR and/or are functionally selective with respect to D$_2$ DAR signaling pathways. One such chemotype discovered this way (compound A) selectively activates the D$_2$ DAR in comparison with other DAR subtypes. We found that compound A exhibits full agonist activity with EC$_{50}$ values ranging from 100 nM – 1 μM using three different functional assays for the D$_2$ DAR: Ca^{2+} mobilization, inhibition of cAMP accumulation, and β-arrestin recruitment. Using β-arrestin recruitment assays to compare with other DARs, we found that compound A has no activity at D1-like DARs (D$_1$ and D$_5$) or on D$_4$ DARs. However, compound A displays either weak partial agonist (<20% of the DA response) or full antagonist activity at D$_3$ DARs. Radioligand binding assays revealed that compound A exhibits Ki values of ~1 μM and ~100 nM for the D$_2$ and D$_3$ DARs, respectively. Interestingly compound A is a full antagonist with no agonist activity on D$_2$ linked GIRK channel activation.

CONCLUSIONS: In summary, compound A is a full and selective agonist at G protein-linked and arrestin mediated D$_2$ DAR assays, however it shows antagonist activity in D$_2$ GIRK channel assays suggesting that it exhibits functionally selective agonist properties. Furthermore, it also functions as a D$_3$ DAR antagonist on all D$_3$ functional assays tested. This is the first known compound that can selectively activate the D$_2$ DAR in the absence of D$_3$ DAR activation.

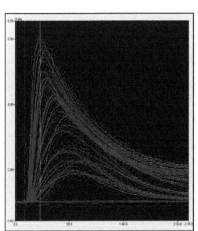

Discovery and Characterization of TAAR1 Agonists, Modulators of Dopaminergic and Serotonergic Neurotransmission with Therapeutic Potential in Neuropsychiatric Disorders

Marius C. Hoener[1]

[1]*F. Hoffmann-La Roche, Basel, Switzerland*

BACKGROUND: Dysregulation of monoaminergic neurotransmission is a hallmark of major neuropsychiatric disorders. The trace amine-associated receptor 1 (TAAR1) is a G protein-coupled receptor activated by trace amines like p-tyramine and β-phenylethylamine, endogenous compounds with structural similarity to biogenic amines.

OBJECTIVES: TAAR1 agonists were developed, characterized and together with TAAR1 transgenic animals used to evaluate TAAR1 as a potential drug target for neuropsychiatric disorders.

METHODS AND RESULTS: By manipulating TAAR1 activity by use of either *Taar1* knock-out animals or selective ligands, we show that TAAR1 modulates dopaminergic, serotonergic and potentially glutamatergic neurotransmission and thus reveal that TAAR1 activation represents a novel therapeutic option for neuropsychiatric disorders. We identified the first potent and selective TAAR1 ligands through a medicinal chemistry program taking advantage of the considerable overlap between the pharmacophore space occupied by TAAR1 ligands and ligands of other biogenic amine receptors. In rodents, activation of TAAR1 by highly potent, selective and pharmacologically distinct compounds blocked psychostimulant-induced hyperactivity and produced a brain activation pattern in phMRI reminiscent of the antipsychotic drug olanzapine, suggesting antipsychotic-like properties. Importantly, TAAR1 agonists did not induce the typical side-effects produced by current antipsychotic drugs, such as catalepsy or weight gain. TAAR1 partial agonism even reduced haloperidol-induced catalepsy and, remarkably, prevented olanzapine from increasing body weight and fat accumulation. Finally, TAAR1 agonists produced pro-cognitive effects (object retrieval paradigm, attentional set-shifting task) as well as antidepressant-like properties (forced swim test, differential reinforcement of low-rate behavior paradigm) in rodent and primate models.

CONCLUSIONS: These data suggest that TAAR1 agonists may provide a novel and differentiated treatment of schizophrenia as compared to current medication standards: TAAR1 agonists may improve not only the positive symptoms but also the negative symptoms and cognitive deficits, without causing adverse effects such as motor impairments or weight gain.

Neurology

David Goldstein and Lee Eiden

Severe depletion of the catecholamine, dopamine (DA), in the striatum (putamen and caudate) is the defining neurochemical characteristic of Parkinson disease (PD). PD also features loss of noradrenergic innervation in the brain and heart. A substantial proportion of the program of the Neurology Theme therefore focused on biomarkers, mechanisms, and treatments of catecholamine deficiency in PD. A pathogenetic concept is emerging based on autotoxicity of cytosolic dopamine. Hastings presented evidence for dopamine toxicity via oxidation to quinones, while Doorn focused on 3,4-dihydroxyphenylacetaldehyde (DOPAL). DA metabolism is channeled through DOPAL via the action of monoamine oxidase (MAO). DOPAL is toxic by at least four mechanisms—protein cross-linking, oxidation to quinones, production of hydroxyl radicals, and oligomerization and precipitation of alpha-synuclein. Two processes efficiently prevent DOPAL from building up in dopaminergic neurons—vesicular sequestration of cytosolic DA via the type 2 vesicular monoamine transporter (VMAT2) and metabolism of DOPAL by aldehyde dehydrogenase (ALDH). The "catecholaldehyde hypothesis" predicts that interference with these processes increases cytosolic DOPAL levels and thereby exerts cytotoxic effects in DA-containing cells. Consistent with the catecholaldehyde hypothesis, Alter et al. reported that VMAT2-deficient mice undergo progressive nigrostriatal degeneration, as well as noradrenergic degeneration, indicating that reduced catecholamine storage is sufficient to produce loss of catecholamine neurons in brain regions other than the nigrostriatal system that are affected in PD. Strong et al. and Wey et al. found that mice with double knockout of the genes encoding ALDHA1A1 and 2 also have neurobehavioral and neuropathologic findings reminiscent of PD. Sossi et al. noted that neuroimaging abnormalities of the nigrostriatal DAergic system are measurable at least 10 years before expected onset of symptoms; an increase in DA turnover appears the most pronounced abnormality, followed by a decrease in binding values of the PET markers of the cell membrane DA transporter and VMAT2. Based on findings in PD patients harboring mutations of LRRK2, the most common genetic risk factor for PD, Sossi et al. proposed that DA turnover might be elevated from birth in these subjects. Finally, Goldstein et al. reported that sympathetic noradrenergic denervation in Lewy body diseases is associated with decreased vesicular uptake of intra-neuronal catecholamines.

Greenamyre et al. presented evidence for bi-directional synuclein-mitochondrial interactions in substantia nigra DA neurons. Better understanding of links between alpha-synuclein and factors determining levels and fates of cytosolic catecholamines should help elucidate mechanisms and incite development of novel treatment and prevention strategies for PD. According to Mosharov et al., alpha-synuclein normally is degraded by chaperone-mediated autophagy (CMA) following transport into the lysosomes. In nigral DA neurons, high intracellular Ca^{2+} levels upregulate cytosolic DA synthesis, and the associated oxyradical stress induces cell defense mechanisms including CMA; however, DA-modified alpha-synuclein blocks uptake of CMA substrates, including alpha-synuclein itself, resulting in synuclein oligomerization to toxic protofibrils that permeabilize synaptic vesicles, leading to further increases in cytosolic DA. The model provides mechanisms for formation of neuromelanin and Lewy bodies and suggests that dysfunction in different metabolic systems may lead to the same pathological changes. Cebrián et al., studying the role of neuroinflammation in neurodegeneration, found that catecholaminergic neurons are more prone than other neurons to express MHC-1 upon IFN-γ challenge, in keeping with a theory that microglial activation or high cytosolic DA triggers catecholamine neuronal antigen display, and the combination of the appropriate antigen and cytotoxic T cells produces the neurodegeneration found in PD.

Kaufmann reviewed norepinephrine (NE) deficiency in PD. Loss of NE seems crucial for both the onset and progression of PD, and data suggest that degeneration of NE-containing neurons in the locus coeruleus, in the pons, precedes degeneration of DA neurons in the substantia nigra. NE deficiency seems to hasten subsequent degeneration of DA neurons. Potential treatments for NE deficiency include inhibitors of the cell membrane NE transporter, presynaptic alpha-2 adrenoceptor blockade, and the NE pro-drug L-dihydroxyphenylserine.

Goldstein et al. reported on cardiac sympathetic denervation, detected by ^{18}F-dopamine PET scanning, as a biomarker of Lewy body diseases. Treglia et al. concurred, based on studies using cardiac ^{123}I-metaiodobenzylguanidine and striatal ^{123}I-Ioflupane scintigraphy. Sharabi et al. reported on CSF biomarkers of central catecholamine deficiency in synucleinopathies. They reported that synucleinopathies feature CSF neurochemical evidence for central DA and NE deficiency. CSF dihydroxyphenylacetic acid seems to provides a sensitive means to identify even early PD.

Khakimova et al. and Nigmatullina et al. modeled the transition from preclinical to clinical PD using a mouse partial denervation model based on different amounts of injected MPTP.

"Presymptomatic" mice had decreased TH protein in axons but not in cell bodies, suggesting decreased axonal transport in the nigrostriatal and tuberoinfundibular DA systems.

Rare cases of familial PD result from mutation of the gene encoding parkin (PARK2). Parkin deficiency seems to increase vulnerability of DA neurons. Liu et al. reported on overexpression of parkin in the nigrostriatal system protecting DA terminals from methamphetamine neurotoxicity. This protection seems related to regulation of the cell membrane DA transporter.

Darvas and Palmiter asked whether DA in the dorsal striatum contributes to behaviors corresponding to cognitive dysfunction in PD. After microinjection of a Cre-recombinase expressing, retrogradely transported viral vector into the dorsal striatum of conditional tyrosine hydroxylase (TH) knock-out mice, the ability to shift strategies (cognitive flexibility), and spatial working memory were decreased. Learning of simple spatial tasks was unaffected, supporting the view that DA in the dorsal striatum contributes to executive function-related behaviors.

Regarding treatment and prevention, Chesselet et al. studied mitochondria-targeted cholesterol-oximes in mice over-expressing wild-type human alpha-synuclein under the Thy1 promoter. The cholesterol oxime, TRO40303, increased transcription of DA-related genes, including TH, LAAAD, DAT, DA2 receptors, and VMAT2, and behavioral analysis supported increased DA function. Schneider et al. reported on a novel treatment method using triple-deuterated levodopa. Because of the deuterium isotope effect, deuterated levodopa may be more potent and less toxic that non-deuterated levodopa.

Compensatory mechanisms in early PD likely postpone onset of clinical symptoms, but at the possible cost of increased oxidative stress. Aluf et al. found that in rats with 50% destruction of nigral DA neurons due to i.c.v. 6-hydroxyDA, microdialysate DA was maintained, but with increased levels of a novel oxidative stress marker, of DA quinone, and of MAO-B. Andersen studied age-related elevations in astrocytic MAO-B that result in selective, progressive loss of nigral DA neurons and decreased locomotor activity.

Modeling Parkinson's Disease in a Mouse Model Overexpressing Astrocytic Monoamine Oxidase B (MAO-B): Towards Exploring Novel Therapeutics

Julie K. Andersen

Buck Institute, USA

BACKGROUND: Several lines of evidence suggest that age-related increases in monoamine oxidase B (MAO-B) may contribute to neuropathology associated with Parkinson's disease (PD). The MAO-B inhibitor deprenyl (selegiline, Eldepryl) is a long-standing antiparkinsonian therapy generally used clinically in concert with the dopamine precursor L-DOPA (Sinamet). Although previous clinical studies have suggested that deprenyl treatment alone is not protective against mortality associated with PD, these studies were targeted to symptomatic patients. By the time PD is symptomatically detected, dopamine loss is usually at least 60% and therefore the lack of MAO-B inhibition in these patients does not negate a role for MAO-B increase in pre-symptomatic dopaminergic loss and disease development. Furthermore, rasagaline (Azilect), a more selective irreversible MAO-B inhibitor assessed in "delayed start" studies to avoid confounding symptomatic effects observed in earlier clinical trials has been reported to delay disease development in patients who initiated drug treatment at the earlier stages consistent with a possible neuroprotective effect and in addition appears to reduce drug-associated dyskinesias when used as an adjunct therapy.

OBJECTIVES: In order to directly evaluate the role of age-related elevations in astrocytic MAO-B levels in PD, we created genetically engineered transgenic mouse lines in which MAO-B levels could be specifically induced within these cells in adult animals (Mallajoysyulla et al., PLOS One, 2008; Siddiqui et al., 2011).

METHODS AND RESULTS: Elevation of astrocytic MAO-B in these animals mimicking that occurring with age and/or disease was found to result in a specific, selective and progressive loss of dopaminergic neurons in the substantia nigra (SN), the same subset of neurons preferentially impacted during progression of the human condition. This was accompanied by other known PD-related alterations including selective decreases in mitochondrial complex I activity, increased levels of mitochondrial oxidative stress, and local microglial activation. These pathological alterations were found to correlate with decreased locomotor activity in these animals. Importantly, pharmacological MAO-B inhibition or antioxidant treatment at earlier stages following MAO-B induction was found to be sufficient to arrest development of subsequent events; this has important implications for early intervention in disease progression. Our data demonstrates that elevation of astrocytic MAO-B results in induction of several phenotypic features of PD suggesting that it could be directly involved in several aspects of disease neuropathology.

CONCLUSIONS: Our inducible astrocytic MAO-B transgenics provide a novel model for exploring pathways involved in inititation or progression of several key features associated with PD pathology that account for cardinal motor symptoms associated with the disorder and for therapeutic testing. Recent pre-clinical therapeutics in the context of this model will be discussed.

3,4-Dihydroxyphenylacetaldehyde is an Autotoxic and Protein-Reactive Intermediate of Dopamine Catabolism

Jonathan A. Doorn, David G. Anderson, Virginia R. Florang, Lydia M.M. Vermeer and Laurie L. Eckert

Department of Pharmaceutical Sciences and Experimental Therapeutics, College of Pharmacy, The University of Iowa, Iowa City, IA

Dopamine (DA) is a neurotransmitter important for coordination of movement as well as other biological functions, and oxidation of the neurotransmitter to toxic species is hypothesized to contribute to the selective neurodegeneration observed in Parkinson's disease (PD). Metabolism of DA involves oxidative deamination via monoamine oxidase yielding 3,4-dihydroxyphenylacetaldehyde (DOPAL) and hydrogen peroxide, and such biotransformation greatly increases the toxicity of DA. DOPAL undergoes detoxication via aldehyde dehydrogenases and aldehyde reductases to acid and alcohol products, respectively. Using dopaminergic cells, we have demonstrated that these carbonyl metabolizing enzymes are highly sensitive to products of oxidative stress (i.e., lipid peroxidation), which are believed to be elevated during PD pathogenesis. Failure of DOPAL biotransformation (i.e., oxidation and compensatory reduction) and the subsequent accumulation of this DA-derived aldehyde is concerning as we have found DOPAL to be a strong electrophile that readily modifies proteins via Lys or Cys. Antioxidants such as N-acetyl Cys and ascorbate were discovered to protect proteins from modification by DOPAL. A proteomic scan of dopaminergic cells treated with DOPAL revealed tyrosine hydroxylase (TH) to be a target of the DA-derived aldehyde. Subsequent analysis demonstrated potent inhibition of TH by DOPAL, involving covalent modification of the enzyme. In addition, we found that DOPAL undergoes auto-oxidation to a quinone, and such oxidation produces superoxide anion. Auto-oxidation was enhanced in the presence of superoxide dismutase, and cyclooxygenase 2 catalyzed oxidation of DOPAL to the quinone. Further analysis of DOPAL oxidation chemistry revealed formation of an initial and transient semi-quinone radical which subsequently yielded a reactive quinone that readily cross-linked proteins. Based on these data, the interaction of DOPAL and oxidative stress is predicted to be a contributing factor to pathologic conditions such as PD.

Dopamine Oxidation and Quinone Modification of Proteins: Mechanisms of Toxicity and Implications for Parkinson's Disease

Teresa G. Hastings, Amanda Mortimer and David Hauser

Pittsburgh Institute for Neurodegenerative Diseases and Department of Neurology,
University of Pittsburgh School of Medicine, Pittsburgh, PA 15260

Oxidative stress and mitochondrial dysfunction have been implicated in the pathogenesis associated with Parkinson's disease (PD). Even though pathology in PD is not restricted to dopamine neurons, it is the dopaminergic neurons of the nigrostriatal pathway that are most susceptible to degeneration in PD. The enhanced vulnerability of this population may be linked to the presence of dopamine itself, where under oxidative conditions, the catechol ring of dopamine will oxidize to form reactive oxygen species and dopamine quinones. The electrophilic quinones react readily with the thiol group of protein cysteinyl residues resulting in covalent modification of proteins with toxic consequences. Dopamine exposure has been shown to be toxic both in cell culture and in vivo, causing selective damage to dopamine neurons that correlated with the extent of intracellular DA oxidation and protein modification. Exposure to dopamine quinone causes mitochondrial dysfunction that is blocked by GSH, suggesting alterations in critical proteins. Proteomic techniques were utilized to identify proteins modified by reactive metabolites of dopamine. Results from our laboratory and others have identified critical mitochondrial proteins as well as familial PD-linked proteins as targets for the dopamine quinone. Selenoproteins that serve critical antioxidant functions within the cell are also likely targets for quinone modification. Glutathione peroxidase 4 (Gpx4) is modified and inactivated following exposure to dopamine quinone. The susceptibility of certain proteins to modification suggests they may play a role in the vulnerability of dopaminergic neurons, as well as the pathophysiology of PD, in general.

This work is supported by a research grant from NIH P01NS059806.

Insights into LRRK2-Mutation Related PD from PET Imaging Studies

V Sossi[1,2], P Agarwal[2], J Mckenzie[2], K Dinelle[1,2], M Schulzer[2], J Aasly[3], Z. Wszolek[5], M Farrer[6] and AJ Stoessl[2,4]

[1]*Department of Physics and Astronomy, University of British Columbia, Vancouver British Columbia, Canada;* [2]*Pacific Parkinson's Research Centre, Vancouver, British Columbia, Canada;* [3]*Department of Neuroscience and Neurology, NTNU, Trondheim, Norway;* [4]*XXX;* [5]*Department of Neuroscience, Mayo Clinic, Jacksonville, Florida, USA;* [6]*Department of Medical Genetics, University of British Columbia, Vancouver British Columbia, Canada*

Mutations in the LRRK2 gene are the most common genetic risk factor for Parkinson's disease (PD) and account for approximately 2% of sporadic PD. Several mutations on LRRK2 have been identified which are characterized by different penetrance, average age of disease onset and disease progression. Imaging studies in asymptomatic mutation carriers can provide information on disease induced neurochemical that precede the onset of clinical symptoms thus providing a unique opportunity to investigate disease origin and disease-induced compensatory changes that delay the onset of clinical symptoms compared to disease initiation. In the last 8 years we have collected multi-tracer longitudinal positron emission tomography (PET) data from over 40 asymptomatic and symptomatic LRRK2 mutation carriers and compared them to the neurochemical phenotype obtained from similar longitudinal studies in idiopathic PD1. Once symptoms occur, the LRRK2-related neurochemical phenotype was found to be identical to that observed in sporadic PD2. Abnormalities of the dopaminergic system are measurable at least 10 years before expected onset of symptoms; an increase in dopamine (DA) turnover appears the most pronounced abnormality, followed by a decrease in binding values of the PET markers of the dopamine transporter (DAT) and vesicular monoamine transporter type 2 (VMAT2) 3. DA synthesis rates appear in the normal range, in keeping with the hypothesis that clinical symptoms only occur when the ability of the system to produce adequate DA levels is impaired. Interestingly, data modeling indicates that DA turnover might be elevated from birth in these subjects, possibly suggesting that alterations in DA release mechanisms might contribute to pathogenesis. Preliminary data suggest that the decreases in DAT and VMAT2 in the asymptomatic mutation carries, when observed, are in keeping with disease progression curves extrapolated from PET data measured in idiopathic PD. Not all subjects exhibited abnormal PET values, which is consistent with the incomplete penetrance of this disease.

References

1. Nandhagopal R, Kuramoto L, Schulzer M, Mak E, Cragg J, Lee CS, et al. Longitudinal progression of sporadic Parkinson's disease: a multi-tracer positron emission tomography study. Brain : a journal of neurology. 2009; 132: 2970–9.

2. Adams JR, van Netten H, Schulzer M, Mak E, McKenzie J, Strongosky A, et al. PET in LRRK2 mutations: comparison to sporadic Parkinson's disease and evidence for presymptomatic compensation. Brain : a journal of neurology. 2005; 128: 2777—85.
3. Sossi V, de la Fuente-Fernandez R, Nandhagopal R, Schulzer M, McKenzie J, Ruth TJ, et al. Dopamine turnover increases in asymptomatic LRRK2 mutations carriers. Movement disorders : official journal of the Movement Disorder Society. 2010; 25(16): 2717—23.

Norepinephrine Deficiency in Parkinson Disease

Horacio Kaufmann

New York University School of Medicine, USA

Parkinson disease is defined biochemically as a deficiency of dopamine in the striatum due to degeneration of dopamine neurons in the pars compacta of the substantia nigra. However, in addition to the reduction of dopamine levels there is extensive pathological, biochemical, and recent imaging evidence indicating reduced levels of norepinephrine both in the brain and in the peripheral autonomic nervous system of patients with Parkinson disease. Interestingly, the norepinephrine deficiency was recognized before the finding of the dopamine deficiency in the brain of patients with Parkinson disease. Experimental evidence indicates that loss of norepinephrine is crucial for both the onset and progression of Parkinson disease. Indeed, data suggest that degeneration of norepinephrine containing neurons in the locus ceruleus in the pontine tegmentum precede degeneration of dopamine neurons in the substantia nigra and that deficiency of norepinephrine hastens the subsequent degeneration of dopamine containing neurons. Moreover, indirect but compelling evidence suggests that norepinephrine might have a neuroprotective effect in the parkinsonian brain. Clinical signs and symptoms attributable to norepinephrine deficiency in patients with Parkinson disease include depression, cognitive impairment and orthostatic hypotension. Pharmacological strategies to reverse norepinephrine depletion might provide symptomatic and potential neuroprotective benefits in patients with Parkinson disease. Pharmacological strategies that can ameliorate or reverse norepinephrine deficiency include norepinephrine transporter inhibitors to increase norepinephrine in the synaptic cleft; presynaptic α-2 adrenergic blockade to increase norepinephrine release in remaining noradrenergic neurons; and the artificial amino acid L-dihydroxyphenylserine (L-DOPS, Droxidopa) that is converted to norepinephrine by decarboxylation. Recently completed large phase III clinical trials of droxidopa showed significant improvement of orthostatic hypotension in patients with Parkinson disease and other neurodegenerative disorders.

Progressive Noradrenergic Degeneration in Mice with Reduced Expression of the Vesicular Monoamine Transporter (VMAT2)

Shawn P. Alter[1], Tonya N. Taylor[2], Patricia Sullivan[3], David S. Goldstein[3] and Gary W. Miller[1]

[1]*Emory University, Atlanta, GA, USA;* [2]*University of Minnesota, Minneapolis, MN, USA;*
[3]*NINDS, NIH, Bethesda, MD, USA*

BACKGROUND: Parkinson's disease (PD) pathology consistently features the preferential destruction of the substantia nigra (SNpc) and locus ceruleus (LC). The vesicular monoamine transporter 2 (VMAT2; Slc18a2) is essential for normal catecholamine storage and release; disruptions in these processes render neurons subject to autotoxicity by catechol oxidation. VMAT2-deficient mice undergo progressive nigrostriatal degeneration, have reduced dopamine (DA) content coupled with increased DA turnover, and display motor and nonmotor symptoms of PD (Caudle, 2007; Taylor, 2009). Here, we report the effects of VMAT2 deficiency on the noradrenergic system.

OBJECTIVES: To determine the effects of aberrant catecholamine storage on noradrenergic signaling and neuronal viability in VMAT2 deficient mice.

METHODS AND RESULTS: Neurochemical analyses show that VMAT2 deficient mice have reductions in brain norepinephrine (NE) (cortex, -54%; striatum, -67%; hippocampus, -46%), as well as increased NE turnover (DHPG:NE, cortex, +91%; striatum, +83%; hippocampus, +71%). Immunohistochemistry and 3H-nisoxetine binding reveal reduced expression of the NE transporter in transgenic mice (-20%). Unbiased stereology reveals severe, progressive LC degeneration. VMAT2-deficient mice exhibit significant reductions in TH immunoreactive cell bodies beginning at 12 months of age, which precedes nigral degeneration at 18 months. From 6-30 months of age, VMAT2 deficient animals undergo 72% LC cell loss (wt: 30%), which is markedly more severe than nigral degeneration (54%).

INTERPRETATION AND CONCLUSIONS: These data describe an animal model of progressive catecholaminergic degeneration. VMAT2-deficient mice have reduced brain NE and DA, as well as increased cytoplasmic catecholamine metabolism. Similar neurochemical deficits have been observed in postmortem PD brains. The severity and staging of the LC and SNpc degeneration are notable, as they are consistent with the progression of catecholaminergic cell death in PD patients. These data demonstrate that reduced catecholamine storage in an *in vivo* model is sufficient to produce progressive degeneration of the catecholaminergic brain regions affected in PD.

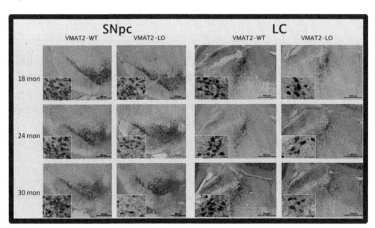

LRRK2 and Risk of PD

Mark Cookson

Laboratory of Neurogenetics, NIA, NIH, USA

Several different mutations in leucine-rich repeat kinase 2 (LRRK2) have been found that are associated with inherited Parkinson's disease (PD). As such, LRRK2 mutations are one of the most common known causes of PD. Pathogenicity confirmed by segregation of mutation with disease for six amino acid changing variants. The N1347H, R1441C and R1441G mutations are found in the ras of complex proteins (ROC) domain, which binds GTP. AY1699C mutation has been reported in the adjacent C-terminal of ROC (COR) domain that likely controls ROC function. Finally, two variants, G2019S and I2020T, are found at adjacent residues in the kinase domain. There are also single amino acids variants that show association with disease, including G2385R in the WD40 domain towards the C-terminus of the protein. Here, I will discuss the ways in which mutations affect protein function and our current best guesses about how this relates to pathogenesis. Of the known mutations, only G2019S increases kinase activity substantially (2−3fold) while others, including the risk factor G2385R, have lower activity and can even overcome the effect of G2019S. Several of the ROC and COR mutations decrease GTPase activity. There are two models for these effects; either GTP binding controls kinase activity, and therefore all mutations would have persistent kinase activity under the right conditions, or kinase mutations are modulatory to GTP binding and hydrolysis, which is therefore critical for pathogenesis. These different views of the LRRK2 molecule and how they should influence our thinking about risk of disease will be discussed.

Synuclein-Mitochondrial Interactions in Substantia Nigra Dopamine Neurons

J.T. Greenamyre[1,2], J.R. Cannon[1], L.H. Sanders[1], V. Tapias[1], Q. Bai[1], L. Volpicelli-Daley[3], V.M. Lee[3] and E.A. Burton[1,2]

[1]*Pittsburgh Institute for Neurodegenerative Diseases and Department of Neurology, University of Pittsburgh, Pittsburgh, USA;* [2]*Geriatric Research, Clinical and Education Center, Pittsburgh VA Healthcare System, Pittsburgh, USA;* [3]*Department of Pathology and Laboratory Medicine, Institute on Aging and Center for Neurodegenerative Disease Research, University of Pennsylvania, Philadelphia, USA*

Multiple lines of evidence strongly implicate both alpha-synuclein and mitochondrial impairment in the pathogenesis of Parkinson disease (PD), but how these mechanisms interact is unclear. We now provide evidence for bidirectional interactions between alpha-synuclein and mitochondria. We have shown that rotenone-induced mitochondrial impairment causes accumulation and aggregation of synuclein in nigral dopamine neurons in vivo. Thus, mitochondrial dysfunction can cause a synucleinopathy. Volpicelli-Daley et al recently reported that application of pre-formed fibrils (PFF) of synuclein to cultured mouse neurons induced Lewy pathology and degenerative changes. We have now applied PFF to cultures of rat mesencephalic neurons and examined mitochondrial respiration and ROS production. At 72 h after application, a time point when there is no degeneration, we find a marked suppression of mitochondrial respiration in intact neurons. Concomitantly, we find that there is a slight increase in mitochondrial ROS generation in treated neurons. Additionally, in PFF-treated cultures, there was a marked enhancement of rotenone-induced ROS production. These results indicate that pathological synuclein has direct effects on mitochondrial function and can also potentiate deleterious effects of mitochondrial impairment. To examine this in an in vivo system, we have employed BAC transgenic rats expressing the E46K mutation of alpha-synuclein, which were created by CJ Li. At 12 months, these rats show altered dopamine metabolism (but no degeneration) and accumulation/aggregation of synuclein in nigral neurons. They also have oxidative damage in substantia nigra. When 6-month-old rats were treated with rotenone, they showed markedly enhanced sensitivity and became parkinsonian faster than wild-type animals; they also have a more extensive loss of dopamine terminals in striatum. These results provide in vivo confirmation that pathogenic forms of synuclein can potentiate the deleterious effects of mitochondrial impairment. To determine whether endogenous synuclein influences susceptibility to mitochondrial dysfunction, we used viral-mediated shRNA to knock down alpha-synuclein unilaterally in nigral dopamine neurons of Lewis rats. Animals were injected in 1 hemisphere with AAV2.GFP and with AAV2.shRNA in the other. Specific, unilateral knockdown of alpha-synuclein was confirmed by in situ hybridization and immunocytochemistry. Rats were exposed to rotenone beginning 3 weeks after AAV2 injection. Unlike control animals, AAV2-injected rats did not become parkinsonian. When they were euthanized after 3 weeks of rotenone treatment, it was found that synuclein knockdown protected the ipsilateral nigrostriatal terminals and dopaminergic cell bodies and their processes in substantia nigra. Thus, endogenous synuclein influences the vulnerability of nigral neurons to mitochondrial impairment. Together, our results indicate that mitochondrial dysfunction can cause a synucleinopathy and, conversely, synuclein can impair mitochondrial function in nigral neurons and potentiate existing mitochondrial dysfunction. Further, modulation of synuclein may have therapeutic potential even in cases of PD caused by environmentally- or genetically-induced mitochondrial defects.

This work was supported by research grants from the United States Department of Veterans' Affairs (1I01BX000548), the National Institutes of Health (1P01NS059806, 1RC1ES018058, 1R01ES020718, 1K99ES019879; 4R00ES019879), and the JPB Foundation.

Mitochondria-Targeted Cholesterol Oximes Increase Dopamine-Related Gene Expression and Behavior in Mice Over-Expressing Alpha-Synuclein, a Model of Pre-Manifest Parkinson's Disease

Marie-Francoise Chesselet[1], Franziska Richter[1], Fuying Gao[1], Sheila M. Fleming[1], Magali Michaud[2], Chunni Zhu[1], G. Coppola[1], Thierry Bordet[2] and Rebecca Pruss[2]

[1]*Department of Neurology, The David Geffen School of Medicine at UCLA, 710 Westwood Plaza, Los Angeles, CA 90095-1769, USA;* [2]*Trophos S.A., Parc Scientifique de Luminy, Case 931, 13288 Marseille Cedex 9, France*

Accumulation of alpha-synuclein and mitochondrial dysfunction are key contributors to the pathophysiology of Parkinson's disease (PD). Mice over-expressing wild-type human alpha-synuclein under the Thy1 promoter (Thy1-aSyn mice; Masliah's line 61) exhibit mitochondrial dysfunction selectively in the striatum and substantia nigra and present progressive behavioral alterations similar to those seen in early stage PD patients. We administered the mitochondria-targeted cholesterol-oximes, TRO19622 and TRO40303 formulated in food pellets at two different dosages, or control food pellets, to male Thy1-aSyn mice and their wild-type littermates from 1 to 4 months of age. Behavior and neuropathology were assessed and tyrosine hydroxylase positive (dopaminergic) neurons were isolated from the substantia nigra by laser capture microdissection. Transcriptome analysis showed dysregulation of 1125 genes in response to alpha-synuclein overexpression, including genes highly implicated in mitochondrial function and oxidative stress (e.g. Bcl-2, Bax, Casp3, Nos2). TRO19622 and TRO40303 normalized about 20% of these genes in transgenic mice. Weighted gene co-expression network analysis (WGCNA) showed a strong effect of TRO40303 on gene expression with enrichment of mitochondrial, cytoprotective and anti-oxidant-response related transcripts. Pathway analysis revealed that TRO40303 increases transcription of dopamine-related genes (Th, Ddc, Gch1, Dat, Vmat2, Drd2, Chnr6a) and behavioral analysis supported increased dopamine function. Neither compound showed any sign of toxicity or adverse effects. This data supports the further development of mitochondrial-targeted cholesterol-oximes for the treatment of PD.

Supported by the Michael J. Fox Foundation, PHS P50 NS38367, and gifts to the UCLA Center for the Study of Parkinson's Disease.

Induction of Chaperone-Mediated Autophagy as a Means to Attenuate Alpha-Synuclein-Mediated Neurotoxicity

Maria Xilouri[1], Oystein Brekk[1], Natalie Landeck[2], Themistoklis Papasilekas[1], Takis Pitichoutis[3], Kostas Vekrellis[1], Zeta Daifoti[3], Deniz Kirik[2] and Leonidas Stefanis[1,4]

[1]*Laboratory of Biology of Neurodegenerative Diseases, Biomedical Research Foundation of the Academy of Athens (BRFAA), Athens, Greece;* [2]*Lund Univeristy, Lund, Sweden;* [3]*Department of Pharmacology, University of Athens Medical School, Athens, Greece;* [4]*Second Department of Neurology, University of Athens Medical School, Athens, Greece*

Aberrant alpha-synuclein (AS) is linked to the pathogenesis of Parkinson's Disease (PD) and other neurodegenerative conditions termed synucleinopathies. One of the critical factors that determine the neurotoxic potential of AS is its levels. Excess of AS leads to PD in humans and to abnormal protein deposition and neurodegeneration in experimental animals. Levels of AS are in part regulated by mechanisms of protein degradation. We and others have provided evidence that Wild Type (WT) AS is degraded in part by the lysosomal process of Chaperone-Mediated Autophagy (CMA) (Cuervo et al., 2004; Vogiatzi et al., 2008). PD-associated mutant forms of AS are not degraded by this process, and in fact impede degradation of other CMA substrates (Cuervo et al., 2004; Xilouri et al., 2009). Furthermore, there is some evidence for a decrease of CMA components in PD brains (Alvarez-Erviti et al., 2010). We reasoned that a possible therapeutic avenue against the toxic effects of AS would be induction of CMA, so as to, on the one hand, accelerate AS clearance, and, on the other, attenuate its toxic effects on lysosomes. To this end, we first created stable SH-SY5Y neuroblastoma cell lines, which express empty vector or the lysosomal transmembrane protein Lamp-2a, the rate-limiting step in the CMA pathway. Lines expressing Lamp-2a showed a significant induction of CMA activity compared to empty vector controls. Steady-state AS levels did not differ, but the half-life of AS was significantly shorter in the Lamp-2a-expressing lines. Significantly, lines expressing Lamp-2a were specifically protected against adenoviral WT AS-mediated neurotoxicity. Similar findings were achieved in rat cortical neuron cultures, where we induced CMA activity by infecting cultures with adenovirus expressing Lamp-2a.

To extend this approach to the in vivo level, we are examining the impact of Lamp-2a overexpression in the rAAV model of nigral WT AS overexpression. We first generated high quality rAAV encoding Lamp-2a and performed intranigral stereotaxic injections. We achieved robust expression of the transgene without any observable toxicity. We then performed similar injections of rAAVs in 3 groups: GFP + Lamp-2a (Group A), GFP + AS (Group B), AS + Lamp-2a (Group C). We are currently analyzing the effects of these injections on the nigrostriatal system, but preliminary results suggest that the neurotoxicity conferred by AS is curtailed by Lamp-2a co-expression. Overall, such data support the notion that induction of CMA may represent a viable therapeutic strategy in PD and other synucleinopathies.

Funding for this work was provided by the PDF and the MJFF

The Link between Neuronal Phenotype and Vulnerability in Parkinson's Disease

James Surmeier

Department of Physiology, Feinberg School of Medicine, Northwestern University, USA

The talk will focus on the principles underlying the hypothesis that neuronal physiological phenotype — how a neuron generates and regulates action potentials — makes a significant contribution to its vulnerability in Parkinson's disease and aging. A cornerstone of this hypothesis is that the maintenance of ionic gradients underlying excitability can pose a significant energetic burden for neurons, particularly those that have sustained residence times at depolarized membrane potentials, broad action potentials, prominent $Ca2+$ entry and modest intrinsic $Ca2+$ buffering capacity. This energetic burden is shouldered in neurons primarily by mitochondria, the sites of cellular respiration. Mitochondrial respiration increases the production of damaging superoxide and other reactive oxygen species (ROS) that have widely been postulated to contribute to cellular aging and Parkinson's disease. Many of the genetic mutations and toxins associated with Parkinson's disease compromise mitochondrial function, providing a mechanistic linkage between known risk factors and cellular physiology that could explain the pattern of pathology in PD. Because much of the mitochondrial burden created by this at-risk phenotype is created by $Ca2+$ entry through L-type voltage-dependent channels for which there are antagonists approved for human use, a neuroprotective strategy to reduce this burden is feasible.

Willie Sutton's Getaway Car and the Pathogenesis of Parkinson Disease

David S. Goldstein, MD PhD

Clinical Neurocardiology Section, CNP/DIR/NINDS/NIH, Bethesda, MD 20892 USA

Willie Sutton was an American bank robber. When asked, "Why do you rob banks?" he reportedly replied, "because that's where the money is!" About pathogenetic mechanisms of Parkinson disease (PD) and related disorders, the "money" is in the catecholamines. If we knew what caused the loss of catecholaminergic neurons, we would have better ideas about how to slow or prevent the disease processes. Here I present a concept about the pathogenesis of PD that addresses four questions: (1) Only a very small percent of neurons are catecholaminergic. What makes them susceptible? (2) How do generalized abnormalities (e.g., mutations, toxins) lead to relatively specific loss of catecholaminergic neurons? (3) Why does alpha-synuclein tend to precipitate in catecholaminergic neurons? And (4) Why is PD mainly a geriatric disease? Catecholamine neurons are like the idling engine of a bank robber's getaway car. Vesicular stores of catecholamines leak continuously into the cytoplasm, where they are "combusted" by spontaneous and enzyme-catalyzed oxidation. The toxic byproduct, dihydroxyphenylacetaldehyde (DOPAL), is detoxified by aldehyde dehydrogenase (ALDH). Having leaky vesicles and active but imperfect recycling back into the vesicles enables rapid initiation of movement and prolonged performance without exhausting releasable stores; but there is a cost, because of ongoing formation of catecholaldehydes. Catecholamine neurons depend on vesicular sequestration and ALDH to limit autotoxicity of cytosolic transmitter, and recent findings indicate increased DOPAL buildup with respect to dopamine in the putamen, decreased vesicular sequestration of intraneuronal catecholamines, and decreased ALDH activity in PD. DOPAL also potently oligomerizes alpha-synuclein. This combination can explain the relatively selective loss of striatal dopaminergic neurons and intra-neuronal synuclein deposition in Lewy body diseases. Aging-relatedness of PD may reflect allostatic load—wear and tear—from genotypic and gene expression abnormalities, increased production of autotoxic compounds, imperfect DOPAL detoxification, synucleinopathy, environmental exposures (e.g., to metals and pesticides), continual episodes of stress-related catecholamine release and reuptake, and time, eventually inciting positive feedback loops and neuronal death.

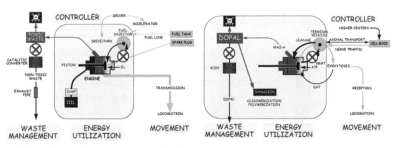

(Left) Deficient fuel recovery and a faulty catalytic converter eventually kill the getaway car engine, with buildup of engine deposits. (Right) Deficient vesicular uptake and decreased ALDH activity eventually kill dopaminergic neurons via DOPAL buildup and alpha-synuclein deposition.

Contributions of Dopamine in the Dorsal Striatum to Behaviors Corresponding to Cognitive Dysfunction in Parkinson's Disease

Martin Darvas[1] and Richard D. Palmiter[1]

[1]*Udall Center - University of Washington, Seattle, Washington, United States*

BACKGROUND: Although the cardinal features of Parkinson's disease (PD) are motor symptoms, PD is now appreciated to cause cognitive impairments that are related to executive function, which is predominantly a domain of the prefrontal cortex. One characteristic feature of PD is the loss of midbrain dopamine projections to the dorsal striatum. The dorsal striatum receives projections from the prefrontal cortex and is part of a multi-synaptic sub-cortical basal ganglia feedback loop. We hypothesize that disruption of striatal dopamine signaling might ultimately impact on the prefrontal cortex and affect executive function.

OBJECTIVES: To examine the effect of restricted and chronic dopamine depletion in the dorsal striatum on executive function related behaviors.

METHODS AND RESULTS: Dopamine depletion was achieved by microinjections of a Cre-recombinase expressing retrograde viral vector into the dorsal striatum of conditional tyrosine hydroxylase knock-out mice. The depletion of dopamine was restricted to the dorsal striatum and ranged from mild (loss of up to 30%) to medium (loss of 30-60%). Two executive function related behaviors, (cognitive flexibility and spatial working memory) and motor coordination were assessed. Depleting dopamine signaling in the dorsal striatum resulted in a dose-dependent deficit in cognitive flexibility, measured as the ability to shift from one strategy to solve a water escape task to a different strategy. Importantly, the ability to utilize either of the two strategies without the shifting condition was not affected by the achieved dopamine depletions.

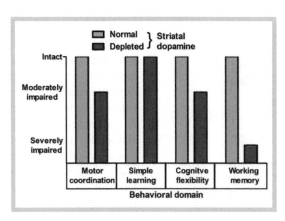

Spatial working memory was equally and severely impaired in mice with both mild and medium dopamine depletion. However, a simple spatial task was not impaired by dopamine depletion in the dorsal striatum, suggesting a deficit specific to the working memory aspect of the task. Both mild and medium dopamine depletions also resulted in a dose-dependant motor deficit on the rotating rotarod.

CONCLUSIONS: Besides the well characterized involvement of dopamine in the prefrontal cortex, our data demonstrate that dopamine in the dorsal striatum is also involved in mediating executive function related behaviors.

Cardiac Sympathetic Denervation: A Biomarker of Lewy Body Diseases

David S. Goldstein, MD PhD

Clinical Neurocardiology Section, CNP/DIR/NINDS/NIH, Bethesda, MD 20892-1620 USA

BACKGROUND: Although the motor features of Parkinson disease (PD) define the diagnosis and constitute the focus of treatment, PD is more than a movement disorder and involves several non-motor manifestations. Dramatic evidence supporting this concept is cardiac sympathetic denervation, which can occur before, about the same time as, or years after clinical onset of the movement disorder. Mechanisms of cardiac sympathetic denervation in PD and other Lewy body diseases remain poorly understood. This study applied a novel combined neuroimaging-neurochemical approach to examine whether Lewy body diseases involve decreased vesicular uptake of intra-neuronal catecholamines. Decreased vesicular uptake would tend to build up cytosolic monoamines, which, via conversion to aldehydes by monoamine oxidase, would be toxic.

METHODS: We used 6-[18F]-dopamine (18F-DA) to track myocardial uptake and retention of catecholamines. Concurrently, the fate of intra-neuronal 18F-DA was followed by assessment of arterial plasma levels of the 18F-DA metabolite 18F-dihydroxyphenylacetic acid (18F-DOPAC). The ratio of myocardial 18F-DA to arterial 18F-DOPAC provided an index of vesicular uptake. Tracer concentrations were measured in patients with PD with or without orthostatic hypotension (PD + OH, PD-No-OH); in patients with pure autonomic failure (PAF, a Lewy body disease without parkinsonism); in patients with multiple system atrophy (MSA, a non-Lewy body synucleinopathy); and in normal controls.

RESULTS: Patients with PD + OH or PAF had decreased vesicular 18F-DA uptake and accelerated 18F-DA loss, compared with MSA and control subjects. PD-No-OH patients could be subtyped into one of these categories based on their initial 18F-DA uptake.

INTERPRETATION: We conclude that sympathetic denervation in Lewy body diseases is associated with decreased vesicular uptake of intra-neuronal catecholamines, suggesting that vesicular monoamine transport is impaired. Vesicular uptake may constitute a novel target for diagnosis, treatment, and prevention.

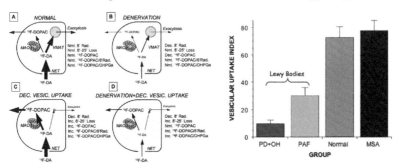

(Left) Concept diagram for the effects of denervation and decreased vesicular sequestration on the uptake and fate of 18F-DA. (A) Normal; (B) reduced sympathetic innervation; (C) reduced VMAT activity; (D) combined decreased sympathetic innervation and reduced VMAT activity. (Right) Vesicular uptake index as a function of diagnostic group.

Reference

Goldstein DS, Holmes C, Kopin IJ, Sharabi Y. Intra-neuronal vesicular uptake of catecholamines is decreased in patients with Lewy body diseases. J Clin Invest 2011;121: 3320–3330.

Cerebrospinal Fluid Biomarkers of Central Catecholamine Deficiency in Parkinson Disease and Other Synucleinopathies

Y. Sharabi, C. Holmes and D.S. Goldstein

Clinical Neurocardiology Section, CNP/DIR/NINDS/NIH, Bethesda, MD 20892-1620 USA

BACKGROUND: Central catecholamine deficiency characterizes alpha-synucleinopathies such as Parkinson disease. We hypothesized that cerebrospinal fluid levels of neuronal metabolites of catecholamines provide neurochemical biomarkers of these disorders.

METHODS: Cerebrospinal fluid catechols were assayed in 146 subjects—108 synucleinopathy patients (34 Parkinson disease, 54 multiple system atrophy, 20 pure autonomic failure) and 38 controls. In 14 patients cerebrospinal fluid was obtained before or within 2 years after the onset of Parkinsonism.

RESULTS: The Parkinson disease, multiple system atrophy, and pure autonomic failure groups all had lower cerebrospinal fluid dihydroxyphenylacetic acid than did the controls (Figure 1). Dihydroxyphenylglycol was also lower in the three synucleinopathies. Dihydroxyphenylacetic acid was lower and dihydroxyphenylglycol higher in Parkinson disease than in pure autonomic failure. Dihydroxyphenylacetic acid was 100% sensitive at 89% specificity in separating patients with recent onset of Parkinsonism from controls but was of no value in differentiating Parkinson disease from multiple system atrophy.

INTERPRETATION: Synucleinopathies feature cerebrospinal fluid neurochemical evidence for central dopamine and norepinephrine deficiency. Parkinson disease and pure autonomic failure involve differential dopaminergic vs. noradrenergic lesions. Cerebrospinal fluid dihydroxyphenylacetic acid seems to provides a sensitive means to identify even early Parkinson disease.

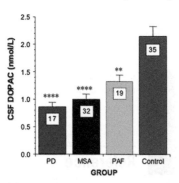 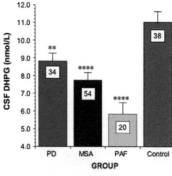

Histograms comparing mean (± SEM) cerebrospinal fluid (CSF) concentrations of (left) dihydroxyphenylacetic acid (DOPAC) and (right) dihydroxyphenylglycol (DHPG) in patient groups with synucleinopathy and control subjects. Abbreviations: PD = Parkinson disease; MSA = multiple system atrophy; PAF = pure autonomic failure. Numbers in each group are shown in white squares. (*) different from control, $p < 0.05$; (**) $p < 0.01$; (****) $p < 0.0001$.

Reference

Goldstein DS, Holmes C, Sharabi Y. Cerebrospinal fluid biomarkers of central catecholamine deficiency in Parkinson disease and other synucleinopathies. Brain 2012 135:1900–1913

Comparison Between Myocardial Sympathetic and Striatal Dopaminergic Imaging in Lewy Body Diseases: "Union is strength!"

Giorgio Treglia[1], Ernesto Cason[2], Alessandro Giordano[1] and Giorgio Fagioli[2]

[1]Nuclear Medicine, Catholic University, Rome, Italy; [2]Nuclear Medicine, Major Hospital, Bologna, Italy

BACKGROUND: Parkinson's disease (PD) and dementia with Lewy bodies (DLB) are overlapping diseases: Lewy body diseases (LBD) has become a general term for these diseases. Both myocardial sympathetic and striatal dopaminergic imaging were used in LBD diagnosis.

OBJECTIVES: To compare myocardial scintigraphy with ^{123}I-MetaIodoBenzylGuanidine (MIBG, a catecholamine analogue) and striatal scintigraphy with ^{123}I-Ioflupane (FP-CIT, a dopamine analogue) in a large patient population with suspected LBD.

METHODS AND RESULTS: 99 patients who performed both methods for differential diagnosis between PD and other parkinsonism (n = 68) or between DLB and other dementia (n = 31) were enrolled. Scintigraphic protocols: A) MIBG scintigraphy: after i.v. injection of 111 MBq of MIBG, early (at 15 min) and delayed (at 240 min) planar images of the chest are obtained; B) FP-CIT scintigraphy: 240 minutes after i.v. injection of 148 MBq of FP-CIT cerebral tomographic images are obtained. Sensitivity (Se), specificity (Sp), accuracy (Ac), positive and negative predictive values (PPV and NPV) of both methods were calculated.

For MIBG scintigraphy, the overall Se, Sp, Ac, PPV and NPV in LBD were 83%, 79%, 82%, 86% and 76% (in differential diagnosis between PD and other parkinsonism 80%, 75%, 78%, 82%, 72%; in differential diagnosis between DLB and other dementia 90%, 91%, 90%, 95%, 83%). For FP-CIT SPECT, the overall Se, Sp, Ac, PPV and NPV in LBD were 93%, 41%, 73%, 71% and 80% (in differential diagnosis between PD and other parkinsonism 95%, 21%, 65%, 63%, 75%; in differential diagnosis between DLB and other dementia 90%, 91%, 90%, 95%, 83%). There was a statistically significant difference between these two methods in patients without LBD (p < 0.05; McNemar's test), but not in patients with LBD (p = 0.1; McNemar's test). The correlation between these methods was not statistically significant in LBD patients.

CONCLUSIONS: LBD usually present both sympathetic and dopaminergic impairment. FP-CIT scintigraphy shows high sensitivity in diagnosing LBD. MIBG scintigraphy may have a complementary role in differential diagnosis between PD and other parkinsonism. These scintigraphic methods show similar diagnostic accuracy in differential diagnosis between DLB and other dementia.

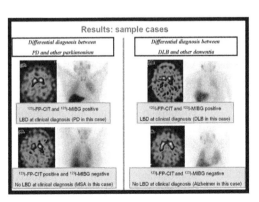

Results: sample cases

Differential diagnosis between PD and other parkinsonism

Differential diagnosis between DLB and other dementia

^{123}I-FP-CIT and ^{123}I-MIBG positive
LBD at clinical diagnosis (PD in this case)

^{123}I-FP-CIT and ^{123}I-MIBG positive
LBD at clinical diagnosis (DLB in this case)

^{123}I-FP-CIT positive and ^{123}I-MIBG negative
No LBD at clinical diagnosis (MSA in this case)

^{123}I-FP-CIT and ^{123}I-MIBG negative
No LBD at clinical diagnosis (Alzheimer in this case)

A Role For Neuronal MHC-I Display in Parkinson's Disease?

C. Cebrián[1], F.A. Zucca[3], L. Zecca[3], J.D. Loike[2] and D. Sulzer[1]

[1]*Departments of Neurology & Psychiatry, Columbia University, New York City, USA;* [2]*Department of Physiology and Cellular Biophysics, Columbia University, New York City, USA;* [3]*Institute of Biomedical Technologies — Italian National Research Council, 20090 Segrate, Milano, Italy.*

BACKGROUND: Parkinson's disease (PD) features a progressive degeneration of ventral midbrain (VM) catecholamine neurons. While the causes of neuronal death are not understood, neuroinflammatory mechanisms are suspected to contribute to a cascade of events leading to chronic degeneration.

OBJECTIVES: The present study aims at analyzing the expression and potential role of MHC-I in catecholamine neurons that degenerate in PD.

METHODS AND RESULTS: We show that 1) dopaminergic (DA) neurons in primary culture can be induced to express MHC-I, 2) cultured catecholamine neurons are more prone to display MHC−I upon IFN-γ challenge than non-catecholamine neurons (see Figure), 3) substantia nigra (SN) DA neurons in culture display MHC-I in response to microglia activated by neuromelanin or alpha-synuclein, substances found extracellularly in postmortem PD brain, 4) in the absence of microglia, chronic exposure to the DA precursor L-DOPA, induced MHC-I in SN DA neurons, 5) in the presence of the appropriate antigen and cytotoxic T cells, MHC-I expressing neurons die, 6) SN DA neuron can present antigen by MHC-I, 7) in human postmortem samples, MHC-I is specifically expressed by SN DA and locus coeruleus (LC) norepinephric neurons, and 8) CD8 + T cells were found in postmortem SN and LC human samples in close proximity to neurons that express MHC-I.

CONCLUSIONS: Our data suggest an innovative theory to explain the etiology of PD. First, specific catecholamine neuronal antigen display can be triggered by microglial activation or high cytosolic DA and, second, the combination of the appropriate antigen and cytotoxic T cells mediate neurodegeneration.

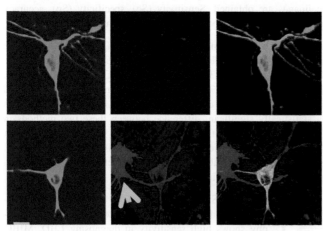

MHC-I immunolabel in postnatally-derived cultured SN DA neurons imaged by confocal microscopy. The upper row shows untreated DA (TH: green) neurons (CON). The bottom row shows MHC-I (red) expressing DA neurons exposed to IFN-γ (100 ng/ml, 72 h). The arrow indicates a MHC-I expressing astrocyte (bottom row). Scale = 30 μm.

Modeling of Preclinical and Clinical Stages of Parkinsonism in MPTP-treated Mice with Focus on Tyrosine Hydroxylase in Survived Nigrostriatal Dopaminergic Neurons

G. Khakimova[1,2], **E. Kozina**[1,2], **V. Kudrin**[2,3], **V. Kucheryanu**[4], **S. Georgieva**[5], **A. Krasnov**[5] and **M. Ugrumov**[1,2]

[1]*Institute of Developmental Biology RAS;* [2]*Institute of Normal Physiology RAMS;* [3]*Institute of Pharmacology RAMS;* [4]*Institute of General Pathology and Pathological Physiology RAMS;* [5]*Institute of Gene Biology RAS, Moscow, Russia*

BACKGROUND: Degeneration of nigrostriatal dopaminergic (DA-ergic) neurons is a key component in pathogenesis of Parkinson's disease (PD). Initial symptoms appear under a threshold loss of DA-ergic neurons (cell bodies) in the substantia nigra (SN) and axons in the striatum.

OBJECTIVES: (i) To model preclinical and early clinical stages of PD in mice with 1-methyl-4-phenyl-1,2,3,6-tetrahydropyridine (MPTP); (ii) To specify the tyrosine hydroxylase (TH) changes in SN and striatum as a whole and TH content in individual cell bodies and axons.

METHODS AND RESULTS: Two-fold MPTP injections at the individual dose of 12 mg/kg provoked the above threshold loss of DA-ergic neurons/DA with no motor dysfunction (presymptomatic stage) whereas 4-fold MPTP injections resulted in a threshold loss of DA-ergic neurons/DA and motor dysfunction (symptomatic stage). At the presymptomatic stage compared to the control: (i) TH protein content decreased in the nigrostriatal system (NSS) and in individual neurons under unchanged level of TH mRNA; (ii) TH protein content decreased in axons remaining unchanged in cell bodies; (iii) TH activity decreased in NSS, to a smaller extent compared to the loss of DA-ergic neurons and TH protein content. At the symptomatic stage compared to presymptomatic one: (i) TH mRNA content decreased in SN whereas TH content increased in individual neurons; (ii) TH content increased in cell bodies remaining unchanged in axons; (iii) TH activity was unchanged in NSS or even increased in individual neurons.

CONCLUSIONS: At both stages of parkinsonism in mice: (i) There was no correlation between the changes in TH transcription and translation rates; (ii) Axonal transport was impaired; (iii) TH activity was unchanged, suggesting that the motor symptoms appearance under the threshold loss of DA was not provoked by the changes in the rate of DA synthesis.

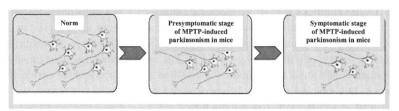

Role of Multiple Factors in Susceptibility of Neurons in Parkinson's Disease

Eugene Mosharov, Elena Kanter, Min Wu, A-M. Cuervo and David Sulzer

Department of Neurology at Columbia University Medical Center, New York, NY 10032

BACKGROUND: Parkinson's disease (PD) is a neurodegenerative disorder that may be caused by genetic and environmental factors. One hypothesis of neurodegeneration in PD postulates that the buildup of cytosolic DA (DA_{cyt}) and its possible interaction with α-synuclein (α-syn) and other PD-related proteins underlie neurotoxicity of neurons affected in PD.

OBJECTIVES: To investigate the synergistic effect of multiple risk factors, including high DA_{cyt} and Ca^{2+} levels, presence of α-syn and inhibition of chaperone-mediated autophagy (CMA). To determine why the cell loss in PD is more severe in SN and locus coeruleus (LC) while other brain DA neuronal populations, including the neighboring ventral tegmental area (VTA) neurons, are relatively spared.

METHODS AND RESULTS: We used SN, VTA and LC neurons from α-syn null mice or their WT littermates. Intracellular patch electrochemistry (IPE) was employed for electrochemical detection of DA_{cyt} in neuronal cell bodies, while the level of toxicity was accessed by counting the density of TH-positive cells following 4 days of treatment with various drugs.

CONCLUSIONS: We demonstrated that neuronal cell death can be induced by the combination of elevated DA_{cyt} and the presence of α-syn. Moreover, SN and LC neurons displayed higher DA_{cyt} and were more susceptible to L-DOPA- and MPP^+-induced neurotoxicities than VTA neurons. This difference was due to higher activity of the $Ca_v1.3$ channels in SN and LC neurons: cytoplacmic Ca^{2+} acted as a 'master switch' that upregulates DA synthesis thereby making cells more susceptible to neurodegeneration. Overall, we show that multiple feedback loops between α-syn, DA, Ca^{2+} and CMA exacerbate cellular stress and lead to selective degeneration of SN neurons.

MODEL OF PD PATHOGENESIS: (1) Physiological function of α-syn is to bind synaptic vesicles in a reversible manner to inhibit exocytosis; (2) α-syn is normally degraded by chaperone-mediated autophagy (CMA) following LAMP2A-mediated transport into the lysosomes. (3) In SN DA neurons, high Ca^{2+} levels upregulate AADC activity, leading to increased DA_{cyt} concentrations and (4) associated oxyradical stress, which induces various cell defense mechanisms, including CMA (5). (6) DA-modified α-syn (7) blocks LAMP2A-mediated uptake of CMA substrates, including α-syn itself. (8) a-Syn oligomerizes to toxic protofibrils, which can (9) bind to and (10) permeabilize synaptic vesicles, leading to further increase in DA_{cyt}. DA stabilizes α-syn protofibrils (11), inhibiting the formation of larger polymers of α-syn (12). The model provides mechanisms for the formation of neuromelanin (NM) and Lewy bodies, as well as suggests that dysfunction in different metabolic systems, i.e. DA metabolism, Ca^{2+} homeostasis, CMA activity and α-syn overexpression or mutation, may lead to similar pathological changes.

Modeling of Presymptomatic and Symptomatic Stages of Parkinsonism in MPTP-treated Mice: Extra-nigrostriatal Manifestations

Razina Nigmatullina[1], Ekaterina Degtyareva[2,3], Vladimir Kudrin[3,4], Tatiana Pronina[2,3], Tatiana Fedoseeva[1], Svetlana Zemskova[1], Gulnara Khakimova[2,3] and Michael Ugrumov[2,3]

[1]Kazan State Medical University, Kazan, Russia; [2]Institute of Developmental Biology RAS; [3]Institute of Normal Physiology RAMS; [4]Institute of Pharmacology RAMS, Moscow, Russia

BACKGROUND: Parkinson's disease (PD) is characterized by the long-lasting preclinical stage and a systemic occurrence of pathological processes besides a degeneration of nigrostriatal dopaminergic (DA-ergic) neurons.

OBJECTIVES: To evaluate functional activity of the tuberoinfundibular DA-ergic system (TIDAS) and heart, and to measure monoamines in plasma of MPTP-treated mice at the presymptomatic (PSS) and early symptomatic stages (ESS) of parkinsonism (Ugrumov et al., 2011).

METHODS AND RESULTS: No changes were observed in TIDAS at the PSS in: (i) the number of DA-ergic neurons and "monoenzymatic" neurons, containing only tyrosine hydroxylase (TH) or aromatic L-amino acid decarboxylase, (ii) TH contents in cell bodies and axons, (iii) DA and norepinephrine (NE) contents. Although the numbers of DA-ergic and monoenzymatic neurons were not changed at ESS: (i) TH content increased in cell bodies and decreased in axons; (ii) DA content dropped whereas TH activity increased, altogether showing functional insufficiency of TIDAS. A failure of the heart was also developed in MPTP-treated mice. Indeed: (i) the weight and contraction force of the left ventricle (LV)were notchanged at PSS whereas a NE stimulating influence on its contraction increased; (ii) the weight and contraction of LV decreased at ESS that was accompanied by a decrease of the NE influence on LV contraction. The MPTP action was also manifested by certain changes in monoamine concentrations in plasma.

CONCLUSIONS: The observed changes in functional activity of TIDAS and heart, and in monoamines concentrations in plasma in MPTP-treated mice at PSS and ESS may serve for the development of preclinical diagnostics of PD.

PARAMETR	PSS MPTP 2x12 mg/kg	ESS MPTP 4x12 mg/kg
TUBEROINFUNDIBULAR DOPAMINERGIC SYSTEM		
Number of DA-ergic neurons	→	→
Number of "monoenzymatic" neurons	→	→
Number of "monoenzymatic" neurons	→	→
TH content in cell bodies	→	↓
TH activity in TIDAS	→	↑
TH content in axons	→	↓
DA contents in TIDAS	→	↓
LEFT VENTRICLES		
Weight	→	↓
Contraction	↓	↓
Time of contraction	↓	↓
Contraction reactivity to NE	↑	↓
NE concentration	↓	↑
Epinephrine concentration	→	→
DA concentration	→	→
Serotonin concentration	↓	↑
BLOOD PLASMA		
NE concentration	→	↑
Epinephrine concentration	→	↑
DA concentration	↓	↑

Reference

UgrumovM.V., KhaindravaV.G., Kozina E.A.et al. Modeling of precliniical and clinical stages of Parkinson's disease in mice / /Neuroscience. 2011. V. 181, P. 175−188.

Basal and Graded 6-OHDA — induced Differences in Striatal Dopamine and its Metabolites in two Mouse Lines with A30P—Mutated Alpha-Synuclein

Marjo Piltonen, Mari Savolainen, Timo T. Myöhänen and Pekka T. Männistö

Faculty of Pharmacy, Division of Pharmacology and Toxicology, P.O. Box 56 (Viikinkaari 5E), FI-00014 University of Helsinki, Finland

BACKGROUND: Point mutations and multiplications of the alpha-synuclein (aSyn) gene are associated with familial Parkinson's disease (PD). The Ala30Pro (A30P)-mutated aSyn has altered biochemical properties, and some mouse lines expressing it exhibit motor impairments at old age, sometimes accompanied by neuropathology.

OBJECTIVES: The aim of this study was to compare the basal levels, and 6-OHDA —induced changes of striatal dopamine (DA) and its metabolites in two A30P-aSyn mouse lines one with inserted human A30P and mouse wild-type protein (mPrPh(A30P)asyn + / +), and the other only carrying the A30P mutation in the mouse aSyn gene (SNCA(A30P) + / +).

METHODS AND RESULTS: The mice received intrastriatal injections of 0.33, 1 or 3 μg of 6-OHDA (A/P + 0.7; M/L + 1.8; D/V -2.7). At 4-5 weeks post-lesion, striata were dissected for a high performance liquid chromatography -analysis. We found that 6-OHDA —treatment affected the striatal DA concentrations similarly in the two mouse lines. The lesioning procedure depleted striatal DA dose-dependently in WT/WT mice, and also in + / + mice with the higher doses of 6-OHDA. However, the lowest dose of 6-OHDA (0.33 μg), caused an elevation in DA in all + / + mice. In mPrPh(A30P)asyn + / + mice, the DOPAC/DA-ratio was higher in the intact and 0.33 μg 6-OHDA groups. Interestingly, in SNCA(A30P) + / + -mice the DOPAC/DA and HVA/DA-ratios were lower than in WT/WT mice.

CONCLUSIONS: Although the overall changes after 6-OHDA were similar in both mouse lines, our results indicate differences in their DAergic systems. The changes are modest, but generally the metabolite levels tend to be elevated in mPrPh(A30P)asyn + / + mice, pointing an increased dopamine turnover. This implies that the previously reported defects in dopamine storage and release in mPrPh(A30P)asyn + / + mice may also result in higher striatal DA and DOPAC concentrations. In SNCA(A30P) + / + mice, the metabolite levels and thus metabolite/DA —ratios tend to be lower. In this case, we hypothesize that the A30P-mutation in aSyn disrupts its normal function as a negative modulator of DAT-trafficking to the plasma membrane, resulting in enhanced release, uptake and perhaps storage. More detailed studies on the dynamics of dopamine release and uptake, as well as the expression of dopaminergic marker proteins are needed in both mouse lines to pin down the mechanisms behind our findings.

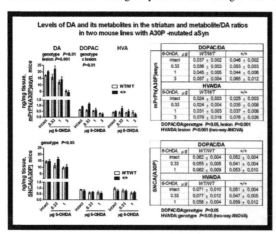

Rasagiline Reduces Striatal Dopamine Release and Extracellular Oxidative Stress in a Rat Model of Early Parkinson's Disease

Y. Aluf[1,2], J. Vaya[2], S. Khatib[2] and J.P.M Finberg[1]

[1]*Department of Molecular Pharmacology, Rappaport Medical Faculty, Technion, Haifa, Israel;* [2]*Laboratory of Oxidative Stress, Migal-Galilee Technology Center, and Tel Hai College, Kiryat Shmona, Israel.*

BACKGROUND: Compensatory mechanisms in early stage Parkinson's disease (PD) postpone the onset of clinical symptoms of the disease, and are a possible cause of oxidative stress (OS). Dopamine (DA) can be metabolized by monoamine oxidase (MAO), or undergo auto-oxidization to form DA quinone, and both these pathways can result in an increase in OS.

OBJECTIVES: To study the effects of partial depletion (50%) of nigrostriatal dopaminergic neurons on striatal extracellular OS, and to explore the effect of selective MAO B inhibition on DA metabolism and redox balance.

METHODS AND RESULTS: Stable 50% destruction of nigral dopaminergic neurons was produced by intracerebroventricular (icv) injection of 6-hydroxydopamine (6-OHDA, 250 μg), and extracellular fluid levels of DA remained at normal levels in microdialysis five weeks later. This compensation was accompanied by increased OS as measured by elevated oxidation of novel marker molecule (LT- scavenges extracellular reactive oxygen species) and DA quinone formation (increased by 44% and 24% respectively). Striatal tissue MAO-B activity was increased by 6-OHDA lesion together with a 45 % increased gliosis. The selective MAO-B inhibitor rasagiline (14 days, 0.05 mg/kg/day, s.c.) did not affect microdialysate DA concentration (DA_{ec}) in sham-operated rats, but decreased extra-cellular OS by 30%. In lesioned rats, rasagiline decreased OS by 50%, and caused a 42% reduction in DA_{ec}, which was reversed by the DA- D2 receptor antagonist sulpiride (10 mg/kg s.c.).

CONCLUSIONS: a) Inhibition of MAO-B by rasagiline (1) in this rat model with striatal gliosis and increased MAO-B activity resulted in an increase in local levels of DA at the presynaptic receptors (2), triggering negative feedback (3) and causing a reduction in DA release to the synaptic cleft (4). b) Rasagiline treatment caused a decrease in intra- and extra-cellular OS (5) which was not explicable by its ability to decrease DA oxidative metabolism.

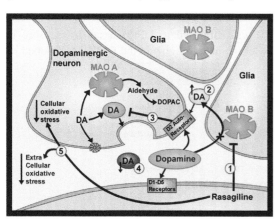

Approach Towards Development of Multifunctional Drugs As an Effective Strategy for Treatment of Parkinson's Disease.

Soumava Santra[1], Liping Xu[1], Mark Johnson[1], Sanjib Gogoi[1], Tamara Antonio[2], Maarten E.A. Reith[2] and Aloke K. Dutta[1]

[1]Pharmaceut. Sci. Wayne State Univ. Detroit, MI; [2]Psychiatry,New York Univ. Schl Med., New York, NY

BACKGROUND: Parkinson's disease (PD) is a progressive neurodegenerative disorder character-ized by degeneration of the nigrostriatal dopaminergic pathway. Both oxidative stress and mito-chondrial dysfunction have been strongly implicated in cell death. In addition, α-synuclein, a presynaptic protein involved in fibrilization, has been implicated in the pathogenesis of PD. It is increasingly evident that for a complex disease such as PD, a drug aimed at one target site will only partially address the therapeutic need of the disease. Thus, it is hypothesized that multi-functional dopamine D2/D3 agonists addressing complex pathogenic factors of PD, will be more effective as symptomatic and disease modifying treatment agents.

OBJECTIVES: In our overall goal to develop such bifunctional/multifunctional drugs as neuropro-tective treatment agents for PD, we designed a novel dopamine D2/D3 agonist molecule, D-512, with potent antioxidant activity and possibly a modulator of alpha synuclein aggregation. In vitro neuroprotection study with dopaminergic MN9D cells was carried out to evaluate effect of D2/D3 agonist D-512 with potent anti-oxidant activity in reversing toxicity of neurotoxic 6-OHDA and MPP + . MN9D cells produce abundant dopamine with high expression of tyrosine hydroxylase enzyme and are considered an ideal in vitro model of PD. Dose dependent effect of D-512 in inhi-bition of Caspase 3 activity was also carried out.

METHODS AND RESULTS: Compound D-512 was subjected to DPPH assay to measure its in vitro antioxidant activity. Compound (-)-**D512** was evaluated *in vivo* in the 6-OHDA-lesioned rat

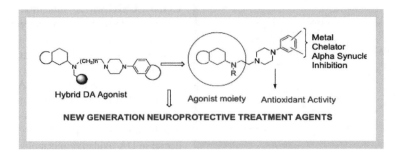

Hybrid DA Agonist   Agonist moiety   Antioxidant Activity

Metal Chelator Alpha Synucle Inhibition

NEW GENERATION NEUROPROTECTIVE TREATMENT AGENTS

rotational model. In neuroprotection study MN9D cells were pre-treated with various concentrations of D-512 or Ropinirole and then, co-treated with 75 μM 6-OHDA or 100 μM MPP + for 24 h. Cell survival was assessed by MTT assay. Caspase 3 assay was carried out to evaluate effect of D-512 on apoptosis in MN9D cells treated with 6-OHDA.

In the anti-oxidant assay compound D-512 was more potent compared to the reference ascorbic acid. D-512 produced the high number of rotations in 6-OHDA lesioned rats, indicating in vivo efficacy in PD animal model. D-512 was able to reverse hypolocomotion in reserpinized (5.0 mg/Kg, s.c.) rats. In cell culture neuroprotection study with dopaminergic MN9D cells, compound (-)-D-512 provided significant dose dependent reduction of toxicity induced by treatment with either 6-OHDA and MPP +, thereby, producing neuroprotection effect. D-512 was able to inhibit apoptotic Caspase 3 signal induced by 6-OHDA in a dose dependent manner, thus, promoting cell survival.

CONCLUSIONS: Novel brain penetrant D2/D3 agonist, D-512, exhibited potent neuroprotection property by rescuing dopaminergic MN9D cells from toxicity of 6-OHDA and MPP +. D-512 dose dependently prevented apoptosis as indicated in inhibition of Caspase 3 activity upon treatment with 6-OHDA. The neuroprotection property of D-512 was independent of dopamine D2/D3 receptor activity, thus, exhibited multifunctional property.

Acknowledgements

We wish to thank Dr. Michael Zigmond (Univ. of Pittsburgh) for the kind gift of MN9D cell lines. This work was supported by NINDS (NS 047198)(AKD)

Triple Deuteration Modifies Pharmacokinetics and Pharmacodynamics of Dopamine in PD Models

Frank Schneider[1], Rudolf-Giesbert Alken[1], Torun Malmlöf[2], Björn Schilström[2] and Torgny H Svensson[2]

[1]*CDRD Berolina AB, Svedala, Sweden;* [2]*Karolinska Institutet, Stockholm, Sweden*

BACKGROUND: L-DOPA, active, or toxic metabolites might contribute to motor and non-motor complications e.g. dyskinesia. Replacement of hydrogen by deuterium might reduce the metabolic breakdown rate of the active substance and the formation of adverse metabolites.

OBJECTIVES: These studies investigate the effects of deuterated L-DOPA + (+ carbidopa) on striatal DA availability and metabolic degradation and the corresponding motor and side effects.

METHODS AND RESULTS: The αββ-triple deuterated BDD-20602 + did not show different peripheral plasma concentration kinetics in rats but generates significantly elevated and sustained striatal DA concentrations and a decreased DOPAC/DA ratio in comparison to L-DOPA + . The increase in striatal norepinephrine (NE) after L-DOPA + is nearly abolished after administration of BDD-20602 + . The equivalent single dose of BDD-20602 + in improvement of motor function was determined to be 60% of L-DOPA + by investigating the dose-response relation of contra-lateral rotations in the 6-OH-DA rat model. This was confirmed by comparing the effects of chronic treatment with 8 mg/kg L-DOPA, 8 mg/kg BDD-20602 and 5 mg/kg BDD-20602 in the cylinder test. BDD-20602 was more effective than L-DOPA at the same dose but did not induce more dyskinesia (abnormal involuntary movements). Furthermore 5 mg/kg BDD-20602 + showed the same motor effect and induced significantly less dyskinesia as 8 mg/kg L-DOPA + . BDD-20602 + (5 mg/kg) did not change the DA/NE ratio in PFC whereas the same L-DOPA + dose doubled the ratio.

CONCLUSIONS: Triple deuteration of L-DOPA reduces the metabolic degradation of the corresponding DA by MAO and DBH and increases the striatal availability. A lower DOPAC/DA ratio indicates that the highly toxic intermediate metabolite DOPAL is diminished as well. Reduced induction of dyskinesia at the same motor effect doses results in a broadening of the therapeutic window. BDD-20602 might open the door for a tuned L-DOPA alternative optimized for unmet needs in PD.

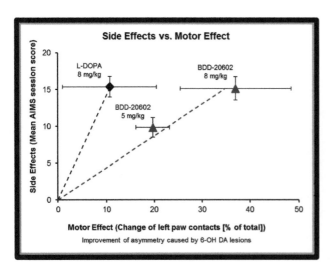

Side Effects vs. Motor Effect

L-DOPA 8 mg/kg

BDD-20602 8 mg/kg

BDD-20602 5 mg/kg

Side Effects (Mean AIMS session score)

Motor Effect (Change of left paw contacts [% of total])

Improvement of asymmetry caused by 6-OH DA lesions

Psychiatry and Psychology

David Goldstein and Lee Eiden

Catecholamines have long been known to contribute to psychological processes such as reward, vigilance, mood, and memory and to major psychiatric disorders such as schizophrenia and depression. In a Plenary lecture, Schultz discussed behavioral functions of explicit dopamine (DA) signals that concern primarily reward value and risk. DA neurons are a component of the brain's reward-encoding system, which extends throughout the striatum, orbitofrontal cortex, and amygdala. The functions of the brain's DA system also include a permissive role in movement and cognitive processes that are deficient in diseases such as Parkinson disease. Midbrain DA neurons are divided into different functional groups. DA neurons in the substantia nigra pars compacta (SNc) and ventral tegmental area (VTA) are key components of the brain's reward system. These neurons are excited by sensory stimuli predicting reward. Based on their response properties it has been thought that DA neurons encode a value-related signal; however, DA neurons do not all encode the value-related signal uniformly. The study by Matsumoto asked: which brain areas provide DA neurons with signals necessary for value- and salience-coding? Experimental findings suggested that the lateral habenula plays an important role as a source of motivational value signal to DA neurons, though the source of their motivational salience signal remains unclear.

Howes et al. reviewed the centrality of DA in the pathogenesis of schizophrenia. Longitudinal studies using 18F-DOPA PET scanning have shown that elevated DA synthesis capacity predates onset of psychosis increases further with development of psychosis, and corresponds with altered glutamatergic function. Consistent with this scenario, Thompson et al. noted imaging studies have indicated that striatal DA release is increased in schizophrenia, predominantly in the precommissural dorsal caudate, and blunted in addiction, mostly in the ventral striatum. The investigators found that in comorbid schizophrenia and substance dependence there is low presynaptic DA release, yet a psychotic reaction to increases in D2 dopamine receptor stimulation, suggesting abnormal post-synaptic D2 function. Grace et al. described bases for abnormal central DA function in schizophrenia. Increased DA neuron activity in the ventral subiculum (VS) causes hyper-responsiveness to phasic stimuli, potentially leading to aberrant attribution of salience. The increased VS activity may in turn result from loss of parvalbumin-positive interneurons in the prefrontal cortex and ventral hippocampus. Yorgason et al. examined effects of early life stress on nucleus accumbens DA signaling. Social isolation (SI) rearing in rats is a model of early life stress that results in disrupted behaviors mimicking depression, anxiety, schizophrenia, or drug addiction. The current study used voltammetric techniques to characterize effects of SI rearing on DA terminal function, including release, uptake, and autoreceptor activity. SI rearing increased DA release, uptake, and paired pulse DA release without affecting autoreceptor activity. Greater DA release may therefore be partly responsible for disrupted DA-related behaviors in isolates.

Ordway et al. provided perspective about roles of norepinephrine (NE) and other monoamines in depression. Evidence that a disorder of monoaminergic transmitters in the brain drives clinical depression remains compelling. NE plays a key role in mediating and modulating behaviors disrupted in depression and is a mediator of the central response to stress, a common precipitator of depression. Two non-monoamine topics of recent intense study are glutamate neurons and glia, and both have associations with NE neurons. NMDA receptor antagonists exert unique and rapid antidepressant activity, and numerous pathologies have indicated either a reduced number of glial cells or reduced glial marker proteins in brains of deceased depressed subjects. Glutamate-NE dysfunction, intertwined with glial deficits, demonstrate the complex multi-neurotransmitter, multi-cellular nature of depression. Drevets et al. reported results of neuroimaging studies of central catecholaminergic function in major depressive disorder (MDD), using fMRI, ^{18}F-deoxyglucose, and ^{11}C-raclopride, including during depressive relapse induced via catecholamine depletion. An observed deficit in central DA release in MDD was consistent with behavioral evidence that depressed subjects show impairment in Pavlovian appetitive conditioning paradigms. These data converge with evidence from preclinical studies to support models in which dysfunction of the mesocorticolimbic DA system contributes to the neural basis of anhedonia and amotivation in MDD. Gold et al. described activated central and peripheral NE secretion in MDD, based on 24-hour CSF sampling and plasma catecholamine levels. CSF and plasma NE, cortisol, and EPI were elevated around-the-clock. These findings argue against the simplest version of the catecholamine hypothesis of major depression, which posits that catecholamine levels are reduced in depression and are responsible for the depressive syndrome.

Aston-Jones et al. reported on upstream modulation of VTA DA neurons by orexin/hypocretin inputs. The orexins (also known as hypocretins) are neuropeptide transmitters in hypothalamic neurons. Many studies have implicated orexin neurons in arousal, but there is also orexin innervation in reward-associated brain regions, including the VTA. Recent data suggest that a lateral septum-lateral hypothalamus orexin-VTA circuit is important in reward-seeking, drug relapse, and addiction.

The locus ceruleus (LC) in the dorsal pons is the main source of NE in the central nervous system. Valentino et al. provided an overview of the LC, stress, opioids, and behavioral flexibility. Their study examined effects of repeated social stress on LC activity in unanesthetized rats during performance of an attentional set shifting task. The results suggested that repeated social stress engages endogenous opioid regulation of the LC, resulting in a state of cellular dependence and increased salience of reward. The two effects might converge to contribute to increased vulnerability to substance abuse in subjects with a history of social stress.

Arnsten pointed out that catecholamines are essential to the higher cognitive functions of the dorsolateral prefrontal cortex (PFC). Recurrent layer III pyramidal cell microcircuits maintain persistent firing through synapses on dendritic spines, and these layer III microcircuits are greatly affected in schizophrenia. Since NE strengthens network connections and improves PFC cognitive functions through alpha-2A adrenoceptor-mediated inhibition of cAMP-HCN and KCNQ channel signaling, the alpha-2A adrenoceptor agonist, guanfacine, is being studied as a possible therapeutic agent.

Meyer, speaking under Theme C-Metabolism, noted that MAO-A is increased in prefrontal and anterior cingulate cortex during major depressive episodes and in high risk states for such episodes, suggesting that preventing the rise in MAO-A level is a potential strategy for preventing onset of illness and maintaining recovery. An emerging neurobiological substrate of depression is the nucleus accumbens, a region that may mediate a diverse range of stressor responses by interfacing limbic, cognitive, and motor circuitry. Corticotropin-releasing factor (CRF), a neuropeptide released in response to acute stressors and other arousing stimuli, acts in the nucleus accumbens to facilitate appetitive processes. Since these processes are, at least in part, mediated by DA, Lemos et al. (abstract not included in these proceedings–see Lemos et al., Nature 490, 402, 2012) hypothesized that CRF acts within the nucleus accumbens to enhance DA transmission. Severe stress was found to disable the capacity of CRF to positively regulate DA, removing CRF's appetitive qualities, leaving a negative perceptual bias. The findings offer a candidate biological substrate for the switch in affect that is central to stress-induced depressive disorders. Yorgason et al. reported that early life stress increases nucleus accumbens DA terminal signaling.

Berridge et al. studied neural mechanisms underlying the cognition-enhancing actions of psychostimulants in attention deficit hyperactivity disorder (ADHD). These drugs improve a variety of prefrontal cortex (PFC)-dependent processes. Methylphenidate elevates extracellular catecholamine levels and enhances neuronal signal processing preferentially within the PFC, via alpha-2 and D1 receptors. Szot et al. studied the relationship between the locus coeruleus (the main source of NE in the brain) and post-traumatic stress disorder (PTSD). Increased central noradrenergic activity contributes to the pathophysiology of PTSD, and excessive responsiveness of the postsynaptic alpha-1 adrenoceptors to NE is an attractive candidate mechanism.

Rewarding Dopamine

Wolfram Schultz

University of Cambridge, United Kingdom

Dr. Schultz will discuss the behavioral functions of explicit dopamine signals which concern primarily reward value and risk and, to a much lesser extent, punishment and physical salience. In coding reward information, the dopamine neurons are a component of the brain's reward system, which includes also the striatum, orbitofrontal cortex and amygdala, whose functions will be briefly mentioned. Despite the well characterized role in reward, the overall functions of the dopamine system extend beyond reward and include a permissive role in a large number of movement and cognitive processes which are deficient in diseases such as Parkinson's.

Modulating Local Circuit Dynamics versus Long-Range Inputs

Christine M. Constantinople and Randy M. Bruno

Dept. of Neuroscience, Columbia University, New York, USA

Mammalian brains generate internal activity independent of environmental stimuli. Internally generated states related to arousal, attention, or reward anticipation may bring about distinct cortical processing modes. To investigate how brain state impacts cortical circuitry, we recorded intracellularly from the same neurons, under anesthesia and subsequent wakefulness, in rat barrel cortex. In every cell examined throughout layers 2–6, wakefulness produced a temporal pattern of synaptic inputs differing markedly from those under anesthesia and natural sleep. Wakefulness had the same effect on both excitatory and inhibitory neurons. Recurring periods of synaptic quiescence, prominent under anesthesia and natural sleep, were abolished by wakefulness, which produced instead a persistently depolarized state. This switch in dynamics was unaffected by elimination of afferent synaptic input from thalamus, suggesting that arousal alters cortical dynamics by neuromodulators acting directly on cortex. Indeed, blockade of noradrenergic, but not cholinergic, pathways induced synaptic quiescence during wakefulness. In contrast, cholinergic antagonists had no effect on on-going dynamics but did alter sensory-evoked activity, likely by modulation of local circuitry mediating feedforward inhibition. We conclude that global brain states can switch local recurrent networks into different regimes via direct neuromodulation. This neuromodulatory mechanism is biophysically and computationally distinct from modulation of specific long-range synaptic projections.

Midbrain Dopamine Neurons are Divided into Different Functional Groups

Masayuki Matsumoto

Kyoto University Japan

Dopamine neurons in the substantia nigra pars compacta (SNc) and the ventral tegmental area (VTA) are well known as key components of the brain's reward system. These neurons are excited by reward and sensory stimuli predicting reward. The excitatory responses increase as the reward value increases. Based on their response property, it has been thought that dopamine neurons encode a value-related signal and to play important roles in reward learning and positive motivation. However, we recently showed that not all dopamine neurons encode the value-related signal uniformly (Matsumoto & Hikosaka, Nature, 2009). We recorded the activity of dopamine neurons in monkeys during a Pavlovian procedure with appetitive and aversive outcomes (liquid rewards and airpuffs directed at the face, respectively). We found that some dopamine neurons were excited by reward-predicting stimuli and inhibited by airpuff-predicting stimuli, as the value hypothesis predicts. However, a greater number of dopamine neurons were excited by both of these stimuli inconsistent with the hypothesis. These neurons are presumed to encode motivational salience which indicates a quantity that is high for both rewarding and aversive events and is low for motivationally neutral events. Notably, the salience-coding neurons were located more dorsolaterally in the SNc, whereas the value-coding neurons were located more ventromedially, some in the VTA. These findings suggest that dopamine neurons are divided into multiple populations encoding distinct signals. Which brain areas provide dopamine neurons with signals necessary for value- and salience-coding? One major candidate is the lateral habenula which has been shown to exert potent negative control over dopamine neurons (Matsumoto & Hikosaka, Nature, 2007). We next recorded the activity of lateral habenula neurons using the same Pavlovian procedure (Matsumoto & Hikosaka, Nature Neuroscience, 2009). We found that most of these neurons were inhibited by reward-predicting stimuli and excited by airpuff-predicting stimuli, indicating that they encode motivational value similar to the value-coding dopamine neurons. The value signals in the lateral habenula would then be transmitted to dopamine neurons by inhibiting them. Our findings suggest that the lateral habenula plays an important role as a source of motivational value signal in dopamine neurons, though the source of their motivational salience signal is still unclear.

Dopamine System Dysregulation by the Hippocampus and the Pathophysiology of Schizophrenia

Anthony A. Grace, Daniel J. Lodge, Kathryn M. Gill and Yijuan Du Holly Moore

Departments of Neuroscience, Psychiatry and Psychology, Center for Neuroscience,
University of Pittsburgh, Pittsburgh, PA 15260, USA

Schizophrenia is believed to arise as a result of a combination of risk factors, including genetic predisposition, birth and gestation difficulties, and environmental impact. We have used a developmental disruption rat model of schizophrenia, in which methyl-azoxymethanol acetate (MAM) is administered to pregnant dams at gestational day 17 and the offspring tested as adults. These offspring recapitulate many of the neuroanatomical, behavioral, and pharmacological features expected of schizophrenia models. We found that MAM treatment correlates with a loss of parvalbumin-stained interneurons in the prefrontal cortex and ventral hippocampus analogous to that reported in schizophrenia. In the ventral hippocampus this corresponds to a decrease in substance P receptor-stained neurons. Given that substance P receptors are markers for parvalbumin interneurons, these data are consistent with a model in which parvalbumin neurons are missing rather than just not expressing this peptide. This interneuron loss corresponds to disrupted gamma rhythms and hyperactivity in the ventral subiculum of the hippocampus. The ventral subiculum hyperactivity, in turn, causes an increase in the number of dopamine neurons firing spontaneously. This increase in dopamine neuron population activity causes the system to be hyper-responsive to phasic stimuli, potentially leading to aberrant attribution of salience. Inactivation of the ventral hippocampus or administration of a hippocampal-specific GABA-A alpha 5 positive allosteric modulator restores dopamine neuron firing to baseline and reverses behavioral hyperactivity to amphetamine. Therefore, we propose that ventral subicular hyperactivity secondary to parvalbumin interneuron loss causes the dopamine system in schizophrenia to be hyper-responsive to stimuli. We propose that the hippocampal damage leading to interneuron loss may be a consequence of an increased responsiveness to stress. Indeed, we found that MAM-treated rats showed less footshock-induced increase in corticosterone and less development of tolerance to footshock-induced corticosterone during the peripubertal period, while showing exacerbated footshock-induced vocalizations. Furthermore, decreasing stressors during this critical period by treating rats with oral diazepam for 10 days peripubertally prevented the exacerbated behavioral response to amphetamine and the increase in DA neuron activity in the adult rat compared to controls. Therefore, an increased sensitivity to the deleterious effects of stress during the peripubertal period may lead to hippocampal damage and interneuron loss, leading to the emergence of psychosis in late adolescence/early adulthood.

The Enduring Centrality of Dopamine in the Pathophysiology of Schizophrenia: In Vivo Evidence from the Prodrome to the First Psychotic Episode

Oliver Howes, Alice Egerton, Arsime Demjaha, Paul Shotbolt, Iris Sommer, Robin Murray, Shitij Kapur and Philip McGuire

King's College London-Institute of Psychiatry, London, SE5 8AF, UK and MRC Clinical Sciences Centre, Imperial College-Hammersmith Hospital, London, W12 0NN, UK

A link between dopamine and schizophrenia has been apparent for many years. However several streams of new evidence have substantially refined understanding of dopamine's role in the pathophysiology of schizophrenia and related psychotic disorders. This evidence derives from in vivo neurochemical imaging studies, findings on the impact of environmental risk factors, and research into people at clinical risk of schizophrenia and from animal models. The findings of over fifty neurochemical imaging findings in schizophrenia show that the major dopaminergic abnormality is presynaptic (see meta-analysis: Howes et al Arch Gen Psych 2012). However, it is not known when dopaminergic dysfunction first occurs in the development of psychotic disorders such as schizophrenia, how it changes with the development of psychosis, or how it relates to treatment resistance. We have addressed these questions using longitudinal studies from the pro-drome to the first psychotic episode, and in studies of patients with established schizophrenia. The following age-matched groups have received [18F]-DOPA PET imaging to measure dopamine synthesis capacity: A) two cohorts of individuals with at risk mental states who are at clinical high risk (CHR) psychosis; B) healthy controls; C) treatment responsive and resistant patients with schizophrenia; D) subjects with long-term subclinical psychotic-like experiences who have never developed a psychotic disorder. The CHR subjects received follow-up to determine who developed psychosis and repeat PET scans to determine change in striatal dopamine synthesis capacity with the onset of the first psychotic episode. Striatal dopamine synthesis capacity was elevated in the CHR cohort at baseline for the whole striatum ($F = 3.7$, $df = 2,42$, $p = 0.035$), and the associativestriatal subdivision ($F = 6.5$, $df = 2,42$, $p = 0.004$), which was confirmed in a second independent cohort. The CHR subjects who developed a psychotic disorder also showed a progressive increase in dopamine synthesis capacity with the development of psychosis($t = 3.01$, $df = 7$, $p = 0.020$). Treatment resistant patients did not show the same pattern of dopaminergic alterations that were seen in the first episode or treatment responsive patients with schizophrenia. Neither the CHR subjects who did not develop a psychotic disorder nor the healthy individuals who reported long-term psychotic-like experiences without developing a psychotic disorder showed an alteration in dopamine synthesis capacity. These findings indicate that elevated dopamine synthesis capacity i) predates the onset of psychosis in people with prodromal symptoms, ii) increases further over time with the development of psychosis; iii) shows an altered relationship with glutamatergic function; but iv) there may be another mechanism in treatment resistance, and in people who report psychotic-like experiences but who do not develop a psychotic disorder.

Striatal Dopamine Release in Schizophrenia Comorbis with Substance Dependence

Judy L. Thompson[1], Nina Urban[1], Mark Slifstein[1], Xiaoyan Xu[1], Lawrence S. Kegeles[1,2], Ragy R. Girgis[1], Yael Beckerman[1], Jill M. Harkavy-Friedman[1], Roberto Gil[1] and Anissa Abi-Dargham[1,2]

[1]Departments of Psychiatry; [2]Radiology, Columbia University College of Physicians and Surgeons, New York, New York, 10032, USA

BACKGROUND: Dopamine plays a role in the pathophysiology of schizophrenia and addiction. Imaging studies have indicated that striatal dopamine release is increased in schizophrenia, predominantly in the precommissural dorsal caudate (preDCA), and blunted in addiction, mostly in the ventral striatum (VST).

OBJECTIVES: Therefore we aimed to measure striatal dopamine release in patients with comorbid schizophrenia and substance dependence.

METHODS AND RESULTS: We used [11C]raclopride PET and an amphetamine challenge to measure baseline dopamine D2-receptor availability (BPND) and its percent change after amphetamine administration (ΔBPND, to index amphetamine-induced dopamine release) in striatal subregions in 11 unmedicated, drug-free patients with both schizophrenia and substance dependence and 15 healthy demographically-matched controls. There were no significant group differences in baseline BPND. Linear mixed modeling using ΔBPND as the dependent variable and striatal ROI as a repeated measure indicated a significant main effect of diagnosis, $F(1, 24) = 8.38$, $p = 008$, with significantly smaller ΔBPND in patients in all striatal subregions (all p values <0.05) except VST. Among patients, change in positive symptoms after amphetamine was significantly associated with ΔBPND in the preDCA (rs = 69, p = 03) and VST (rs = 64, p = 05).

CONCLUSIONS: Patients with comorbid schizophrenia and substance dependence showed significant blunting of striatal dopamine release, in contrast to what has been found in schizophrenia without substance dependence. Despite this blunting, dopamine release was associated with the transient amphetamine-induced positive-symptom change, as observed in schizophrenia. This is the first description of a group of patients with schizophrenia who display low presynaptic dopamine release, yet show a psychotic reaction to increases in D2 stimulation, suggesting abnormal post-synaptic D2 function.

Neuroimaging Studies of Central Catecholaminergic Function in Depression

Wayne C. Drevets and Chantal Martin-Soelch

Laureate Institute for Brain Research, Tulsa, OK, USA and Department of Psychiatry and Psychotherapy, University Hospital Zurich, Switzerland

The mesocorticolimbic circuits that modulate the neural processing of reward and behavioral incentive are implicated in the pathophysiology of major depressive disorder (MDD) both by preclinical data obtained in rodent models of depression and by pharmacological and neuroimaging data acquired in human mood disorders. This presentation reviews the abnormal patterns of neural activity and dynamic neurotransmitter function evident in these circuits in MDD subjects as they perform reward-processing tasks under experimental conditions putatively associated with catecholamine depletion or dopamine release. Using fMRI the neurophysiological correlates of reward processing were compared between currently-depressed MDD subjects (dMDD) and healthy controls as they performed a monetary incentive delay (MID) task. Effects of dopaminergic transmission and mood on the relevant functional anatomy and on reward processing deficits seen on the MID task in dMDD cases were investigated using both behavioral studies and PET-[F-18]flurodeoxygluose glucose metabolic imaging in remitted subjects with MDD (rMDD) during depressive relapse induced via catecholamine depletion. In separate studies, PET-[C-11]raclopride imaging was performed as subjects performed a slot machine task to compare the magnitude of dopamine release in response to unexpected monetary reward between dMDD subjects and healthy controls. While performing the MID task, dMDD subjects showed altered activity in the orbital prefrontal cortex (OFC), hippocampus, amygdala and accumbens in anticipation of initiating behavioral responses to acquire rewards or avoid losses. These abnormalities were associated with impaired modulation of behavioral performance by varying incentive levels. Under catecholamine depletion rMDD subjects, but not controls, develop this same behavioral pattern on the MID task, anhedonia, depressed-mood, and metabolic abnormalities in OFC and accumbens. The dMDD subjects also showed abnormally reduced dopamine release during reward in the ventral striatum. Depressed MDD subjects showed altered cortocolimbic activity in the OFC, hippocampus, amygdala and accumbens, as they anticipated initiating behavioral responses aimed at acquiring rewards or avoiding losses while performing the MID task. These physiological abnormalities were associated with impaired modulation of the behavioral response to changing incentive levels. Currently remitted subjects with MDD showed an abnormal diathesis to develop both this same behavioral pattern on the MID task and abnormal glucose metabolism in the OFC and accumbens area under catecholamine depletion, compatible with evidence that such subjects show reductions in the CSF homovanillic acid levels. Converging with these data, depressed patients also showed abnormally reduced DA release in response to unexpected reward. Based upon preclinical evidence that DA release in response to unexpected reward serves as a teaching signal during the reward learning process, the deficit in central DA release in MDD is consistent with behavioral evidence that depressed subjects are impaired in Pavlovian appetitive conditioning paradigms. These data converge with evidence from preclinical studies to support models in which dysfunction of the mesocorticolimbic dopaminergic system contributes to the neural basis of anhedonia and amotivation in MDD.

Central and Peripheral Norepinephrine Secretion in Major Depression is Activated, as Assessed by 24 hour CSF and Plasma Sampling

Philip W. Gold[1], Murray Esler[2], Gavin Lambert[2], Courtney Holmes[3], David Goldstein[3] and Mitchel Kling[4]

[1]*NIMH Intramural Research Program, Bethesda, Maryland, USA;* [2]*Baker Heart Research Institute, Melbourne, Victoria, Australia;* [3]*NINDS, Bethesda, Maryland, USA;* [4]*University of Pennsylvania, Philadelphia, Pennsylvania, USA*

INTRODUCTION: The classic catecholamine hypothesis of major depression is that catecholamine levels are significantly reduced in depression and are responsible for the depressive syndrome. To further test this hypothesis, we report here a study of around-the-clock levels of CSF and plasma norepinephrine, epinephrine, and cortisol in 12 medication-free severely depressed patients with melancholic depression before and after ECT. We also assessed NE spillover in a separate group of medication-free patients.

METHODS: In the first study, we recruited 12 patients with severe melancholic depression and 12 matched controls. Patients were medication free for one month prior to study. A lumbar catheter was placed at 2000 and CSF was exfused at a constant rate of 1cc per hour for 30 hours. Plasma was drawn hourly from the antecubital catheter as well. CSF and plasma samples were measured for norepinephrine (NE), cortisol, epinephrine (EPI), and DHPG. Corticotropin releasing hormone

(CRH) in CSF was also measured. Three weeks after ECT-induced recovery from depression, all studies were repeated. During each night of the study, sleep was monitored by standard EEG. In the second study, 12 less severely medication-free depressed patients and twelve controls each underwent a study of NE spillover into arterial plasma at baseline, in response to a video game, and after infusion of yohimbine. In the third study, we recruited 15 medication-free patients with major depression, 15 patients with bipolar disorder, 15 with MDD, and 45 controls.

RESULTS: CSF and plasma NE, cortisol, and EPI were significantly elevated around-the-clock. ($p < .02$) The diurnal variations of plasma and CSF NE and cortisol were virtually identical, and each correlated with the other as closely as did plasma ACTH and cortisol. Plasma ACTH levels were similar in patients and controls. The mean levels of plasma NE and cortisol were also highly correlated. Following ECT, CSF NE, plasma NE, plasma cortisol, and plasma EPI all fell to levels similar to those in controls ($p < .01$). CSF CRH levels were similar those in controls during the depressed state and were lower than in controls after ECT-induced remission. In the second study, we found significant elevations in NE spillover at baseline, in response to a video game and in response to yohimbine ($p < ,02$). In the third study, despite similar plasma NE levels in MDD patients and those with BD, plasma DHPG was significantly higher in the MDD patients. than in patients with BD disorder or controls ($p < .01$). For the overall group, there was a strong trend for increased plasma NE in the entire patient group compared to controls ($P < .056$).

CONCLUSIONS: Patients with major depression are in a central and peripheral hypernoradrenergic state with increased sympathetic nervous system activity, all reversible by electroconvulsive therapy. CSF and plasma NE and cortisol have diurnal variations that are virtually identical, peaking at 9AM, a time associated with maximal vulnerability to myocardial infarction and death. A significant decrease in DHPG levels in patients with BD compared to those with MDD, despite similar NE levels, suggest an alteration in endogenous MAOI activity between the groups.

Norepinephrine: A Broken Spoke on the Wheel of Depression

Gregory A. Ordway[1], Attila Szebeni[1], Michelle J. Chandley[1], Craig A. Stockmeier[2], Jessica Crawford[1] and Katalin Szebeni[1]

[1]Department of Biomedical Sciences, James H. Quillen College of Medicine, East Tennessee State University, Johnson City, TN, USA; [2]Department of Psychiatry and Human Behavior, University of Mississippi Medical Center, Jackson, MS, USA

It has been over 60 years since the first modern antidepressant was discovered. During this period, billions of public and private dollars have been spent to develop better treatments for depression and to determine the biological basis of depression. As of 2012, our most popular antidepressants are drugs that mimic the efficacy and noradrenergic and/or serotoninergic effects of the first antidepressants. In contrast to this fact, many recent published theories regarding the biological basis of depression scarcely mention catecholamines and serotonin. This disconnection partly results from the fact that many patients do not respond or respond incompletely to modern antidepressants. Hence, there remains the belief that the biogenic monoamines are not the "final common pathway" and that some other pathognomonic transmitter, growth factor, transcription factor, cell, etc. is yet to be discovered. Despite this search for a possibly mythical final common pathway, the accumulated research on monoamines provides very powerful evidence that a disorder of these transmitters plays a role in the brain pathology of depression. Of these monoamine neurotransmitters, norepinephrine plays a central role in mediating and modulating behaviors that are disrupted in depression, and also is a key mediator of the central response to stress, a common precipitator of depression. Two non-monoamine areas of recent intense study in the field of depression are glutamate neurons and glia, both of which have unique and intimate associations with noradrenergic neurons. Elevations of glutamate in depressed subjects have been demonstrated, and NMDA receptor antagonists demonstrate unique and rapid antidepressant activity. With regard to glia, numerous pathologies have been described with a consistent theme being a dysfunction of glia, as evident by either a reduced number of glia cells or by reduced glial marker proteins in brains of deceased depressed subjects. Recent evidence from my laboratory demonstrates that astrocytes in the immediate vicinity of pontine noradrenergic neurons are unhealthy in depression. In particular, these glial cells demonstrate reduced levels of expression of growth factor and glutamate transporter genes. Glutamate is a stress-sensitive excitatory input to noradrenergic neurons in the pontine brainstem. Hence, reduced astrocytic uptake of glutamate would be expected to increase the potential deleterious effects of stress-activated glutamatergic excitation of central noradrenergic neurons. Elevated expression levels of multiple glutamate receptor genes specifically in noradrenergic neurons in depressed subjects provides further evidence of a glutamate-norepinephrine dysfunction in depression, one likely intertwined with glia deficits. Collectively, these findings further demonstrate the complex multi-neurotransmitter/multi-cellular nature of depression. Pharmacological agents or drug combinations that target multiple pathologies of depression may increase the efficacy of pharmacotherapy of depression and may result in a greater percentage of treatment successes.

PET Imaging of Serotonin Transmission in Monkeys: Effects of Maternal Separation and Long-Term Fluoxetine Treatment during Adolescence

Saurav Shrestha[1,2], Jeih-San Liow[1], Robert Gladding[1], Victor W. Pike[1], Pam L. Noble[1], James T. Winslow[1], Stephen J. Suomi[1], Per Svenningsson[2], Eric E. Nelson[1], Daniel S. Pine[1] and Robert B. Innis[1]

[1]*Intramural research program of National Institute of Mental Health, National Institutes of Health, Bethesda, Maryland, USA;* [2]*Translational Neuropharmacology, Center for Molecular Medicine, Karolinska Hospital, Stockholm, Sweden*

BACKGROUND: Major depressive disorder (MDD) is a serious disorder that often begins following stress during adolescence. Selective serotonin reuptake inhibitors (SSRIs) are a common treatment for both adolescent and adult MDD. While MDD's early onset and available efficacy data support use of SSRIs in adolescents, concerns about safety have arisen, based on associations with suicidal behavior in adolescents, coupled with minimal data on long-term effects on the developing brain. This study uses an established non-human primate model to examine the developmental impact of SSRI exposure.

METHODS: This study was performed in 32 male rhesus macaques, divided into four conditions. One group of 16 randomly-selected monkeys underwent maternal-separation. The other group of randomly-selected 16 monkeys underwent normal-rearing. Half of each group was in turn randomly exposed to either chronic fluoxetine or placebo during the third year of life. Thus, two experimental manipulations were used to generate the four groups of eight monkeys. Chronic fluoxetine treatment began at 2-year of age for one year. One to two years post-washout, monkeys (average age of 5) were scanned using two PET radioligands: 1) [11C]DASB for serotonin transporter (SERT); and 2) [11C]RWAY, a 5-HT1A receptor antagonist. Receptor densities for SERT and 5-HT1A were measured in neocortex, hippocampus and dorsal raphe nucleus.

RESULTS: A significant two-way, rearing-by-treatment interactions emerged for only SERT whereas only main effect for either rearing or treatment was seen for 5-HT1A receptor. A significant global decrease was seen in SERT binding in maternal-separated compared to normal-reared monkeys. A significant increase in 5-HT1A receptor binding was seen in only raphe in maternal-separated compared to normal-reared monkeys. Lastly, SERT but not 5-HT1A receptor binding was reversed in maternal-separated monkeys that received fluoxetine treatment.

CONCLUSIONS: To our knowledge, this is the first study in young adolescent monkeys showing persistent effects of fluoxetine treatment in raphe despite being several years after washout. Our study demonstrates that both SERT and 5-HT1A receptor are altered in maternal-separated monkeys, and chronic flouxetine treatment may normalize deficits in SERT density that is persistent more than one year after medication discontinuation. Our study highlights the importance of environment and treatment during development.

Opposing Regulation of Norepinephrine and Dopamine in Rat Limbic Brain Regions in Response to Aversive Stimuli

Jinwoo Park[1], Elizabeth S. Bucher[2] and R.Mark Wightman[2,3]

[1]*Department of Biotechnical and Clinical Laboratory Sciences, University at Buffalo-SUNY, Buffalo, New York, United States;* [2]*Department of Chemistry, University of North Carolina, Chapel Hill, North Carolina, United States;* [3]*Neuroscience Center, University of North Carolina, Chapel Hill, North Carolina, United States*

BACKGROUND: Norepinephrine and dopamine in limbic forebrain structures regulate and modulate the central pathways of behavioral and physiological response during reward and stress. We investigated and compared different characteristic features of norepinephrine and dopamine neurotransmission in two rat limbic structures, the bed nucleus of the stria terminalis (BNST) and the nucleus accumbens (NAc) in response to aversive stimuli.

OBJECTIVES: This study will delineate the correlation of the dynamic changes in the two catecholamines in response to the acute aversion and after its termination.

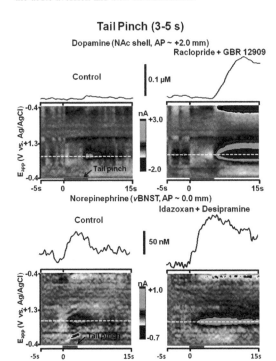

Tail Pinch (3-5 s)

Dopamine (NAc shell, AP ~ +2.0 mm)

Raclopride + GBR 12909

Control

0.1 µM

-0.4
+1.3

nA
+3.0
-2.0

Tail pinch

-5s   0   15s   -5s   0   15s

Norepinephrine (vBNST, AP ~ 0.0 mm)

Idazoxan + Desipramine

Control

50 nM

-0.4
+1.3

nA
+1.0
-0.7

Tail pinch

-5s   0   15s   -5s   0   15s

METHODS AND RESULTS: Fast-scan cyclic voltammetry at carbon-fiber microelectrodes was employed to study norepinephrine and dopamine signaling in the subregions of the BNST and NAc in response to the aversive stimuli. I. *Tail Pinch:* Extracellular norepinephrine and dopamine levels were respectively recorded in the ventral (*v*)BNST and NAc shell of anesthetized rats before, during and after acute tail pinch (3–5 s) with maximal pressure. Our results showed that extracellular norepinephrine in the *v*BNST increased at the onset of the tail pinch and returned to pre-stimulus concentrations after the tail pinch in an uniform manner. In contrast, dopamine transmission in the NAc shell was inhibited during the aversive stimuli. II. *White Noise & Light Stimuli:* After recovery from the stereotactic surgery, changes of extracellular catecholamine concentration of the naïve rats were recorded before, during and after the noise (65 dB) & light (10 lux). Noradrenergic signaling in the *v*BNST was activated during the stimuli and then decreased after the stimuli. In contrast, dopaminergic signaling in the dorsolateral (*dl*)BNST was inhibited during the stimuli and then increased just after the termination of the aversive stimuli.

CONCLUSIONS: The reciprocal relationship of the two catecholamines, coupled with their different time courses, may provide integrative processing of aversive sensory inputs to influence response outputs appropriate for adaptation and survival.

The Locus Coeruleus, Stress, Opioids and Behavioral Flexibility

Rita J. Valentino[1], Nayla Chaijale[1], Kevin Snyder[2], Wei-Wen Wang[3] and Andre Curtis[1]

[1]*The Children's Hospital of Philadelphia, Philadelphia, PA;* [2]*University of Pennsylvania, Philadelphia, PA;*
[3]*Institute of Psychology, Beijing, China*

BACKGROUND: The locus coeruleus (LC)-norepinephrine system is co-regulated during stress by corticotropin-releasing factor (CRF) and endogenous opioids in an opposing manner. CRF increases LC tonic discharge rate and attenuates LC sensory responses. In contrast, μ-OR activation selectively decreases LC tonic activity resulting in an enhanced signal-to-noise ratio. The acute effects of CRF on LC activity would be predicted to promote cognitive flexibility. Consistent with this, intra-LC CRF facilitated extradimensional set shifting.

OBJECTIVES: This study examined the effects of repeated social stress on LC activity in unanesthetized rats during performance in an attentional set shifting task.

METHODS AND RESULTS: LC activity was recorded in unanesthetized rats. Rats were exposed to resident-intruder stress or were put in novel cages (30 min/day; 5 days). LC discharge rates were lower in rats exposed to social stress compared to controls when recorded 2 days after the last social stress. Naloxone (2 mg/kg, ip) increased LC discharge rates of stressed but not control rats at both 2 and 10 days after social stress, suggesting an enduring regulation of LC activity by endogenous opioids following social stress. Naloxone elicited signs of mild opiate withdrawal in stressed, but not control rats at 2 days after the last stress. Stressed rats performed better in the intradimensional shift (IDS) part of the set shifting task compared to controls. LC recording during IDS revealed an increase in reward-related LC activity in stressed, but not control rats, suggesting an enhanced salience of the reward.

CONCLUSIONS: Together, the results suggest that repeated social stress engages an endogenous opioid regulation of the LC that results in a state of cellular dependence and increases the salience of reward. These two effects may converge to contribute to an increased vulnerability to substance abuse in subjects with a history of social stress.

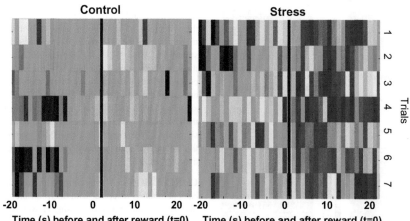

Measurements of Dopaminergic Function in the Rat Brain Using [^{18}F]FDOPA PET and Microdialysis

Matthew D. Walker[1], Katherine Dinelle[1], Rick Kornelsen[1], Siobhan McCormick[1], Chenoa Mah[1], Anna Lee[1], James E. Holden[2], Matthew J. Farrer[1], A Jon Stoessl[1] and Vesna Sossi[1]

[1]University of British Columbia, Vancouver, British Columbia, Canada; [2]University of Wisconsin, Madison, Wisconsin, USA

BACKGROUND: [^{18}F]fluoro-3,4-dihydroxyphenyl-L-alanine ([^{18}F]FDOPA) positron emission tomography (PET) allows non-invasive measurement of dopaminergic function in primates. Prior studies in rodents largely failed, due to more rapid peripheral metabolism.

OBJECTIVES: We demonstrate that [^{18}F]FDOPA PET can measure pre-synaptic dopaminergic function in the rat.

METHODS AND RESULTS: Eight normal control rats underwent microdialysis; tolcapone or entacapone was administered (40 mg/kg, i.p.) Tolcapone, but not entacapone, consistently reduced homovanillic acid (HVA) and increased 3,4-dihydroxyphenylacetic acid (DOPAC) concentrations in striatal dialysates, confirming central inhibition of catechol-O-methyltransferase (COMT). Eleven unilaterally 6-hydroxydopamine (6-OHDA) lesioned and four normal control rats underwent [^{18}F]FDOPA PET. Inhibitors of COMT (entacapone or tolcapone, 40 mg/kg, $t = -90$ min) and aromatic L-amino acid decarboxylase (benserazide, 10 mg/kg, $t = -30$ min) were given i.p. before the 3-hour duration [^{18}F]FDOPA (i.v., $t = 0$) PET scan. In most cases (> 80 %) there was clear striatal radiotracer accumulation; the remaining scans had far less brain uptake and were discarded. [^{18}F]FDOPA uptake (k_{ref}) and radiotracer washout (k_{loss}) were modeled. For normal control rats (with entacapone), striatal estimates were $k_{ref} = 0.037 \pm 0.005$ min^{-1}, the effective distribution volume ratio (*EDVR*) was 1.07 ± 0.22 and $k_{loss} = 0.024 \pm 0.003$ min^{-1}. In lesioned animals, the unilateral striatal denervation measured by [^{11}C]-(+)-dihydrotetrabenazine (DTBZ) PET ranged from $14 - 92$%. The *EDVR* from [^{18}F]FDOPA was reduced in the lesioned striatum; this reduction correlated ($r = 0.9$) with the [^{11}C]DTBZ binding potential (BP_{ND}). Tolcapone increased the *EDVR* (22%median increase c.f. entacapone, $p < 0.01$) by reducing radiotracer washout. Dopamine metabolite levels in the lesioned striatum, measured by microdialysis, were proportional to DTBZ $BP_{ND}(r = 0.9, p < 0.01)$. Recent data suggest improved reliability of [^{18}F] FDOPA PET by over-night fasting.

CONCLUSIONS: Measurements of pre-synaptic function were made using [^{18}F]FDOPA PET. These measurements were sensitive to the effects of 6-OHDA lesioning and the pharmacological inhibition of central COMT. PET measures were correlated with those from microdialysis.

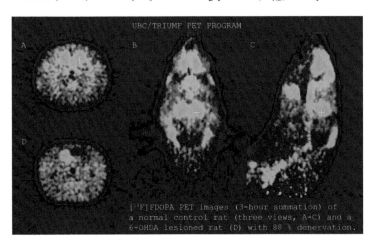

[^{18}F]FDOPA PET images (3-hour summation) of a normal control rat (three views, A-C) and a 6-OHDA lesioned rat (D) with 88 % denervation.

Evaluation of [^{11}C]ORM-13070 as a PET tracer for α$_{2C}$-adrenoceptors in the human brain

Mika Scheinin[1,2]**, Mika M Hirvonen**[2]**, Jarkko Johansson**[3]**, Jukka Kemppainen**[3]**, Jussi Lehto**[1,2]**, Zsofia Lovro**[1,2]**, Pauliina Luoto**[3]**, Vesa Oikonen**[3]**, Tarja Naukkarinen**[4]**, Juha Rouru**[4]**, Jukka Sallinen**[4]**, Harry Scheinin**[3]**, Lauri Vuorilehto**[2]**, Sjoerd J Finnema**[5]**, Christer Halldin**[5] **and Juha O Rinne**[3]

[1]*Department of Pharmacology, Drug Development and Therapeutics, University of Turku, Turku, Finland;* [2]*Clinical Research Services Turku CRST, Turku, Finland;* [3]*Turku PET Centre, University of Turku and Turku University Hospital, Turku, Finland;* [4]*Orion Corporation Orion Pharma, Turku, Finland;* [5]*Karolinska Institutet, Stockholm, Sweden*

BACKGROUND: The α$_{2C}$-adrenoceptor (AR) antagonist radioligand [^{11}C]ORM-13070 is a novel α$_{2C}$-AR PET imaging agent. PET imaging in non-human primates has indicated that its binding is sensitive to amphetamine-induced changes in extracellular norepinephrine (NE) levels.

OBJECTIVES: This study aimed to determine the maximal extent of inhibition of striatal [^{11}C]ORM-13070 uptake by the α$_2$-AR antagonist atipamezole in order to validate [^{11}C]ORM-13070 for occupancy studies. A secondary objective was to determine whether striatal [^{11}C]ORM-13070 uptake could be reduced by physiological and pharmacological NE challenges.

METHODS: 8 healthy male volunteers were investigated. PET imaging was performed after different doses (up to 450 μg/kg) of atipamezole and four different NE challenges. The NE challenges were expected to increase NE release in the brain (cold pressor test, insulin-induced hypoglycaemia and ketamine infusion) or to inhibit its neuronal reuptake (atomoxetine). Each subject underwent one baseline PET scan, one atipamezole scan and two scans with two different NE challenges.

RESULTS: Atipamezole inhibited striatal tracer uptake in a dose-dependent fashion with an average E$_{max}$ of 78% and an ED$_{50}$ of 13 μg/kg in the caudate nucleus (EC$_{50}$ in blood, 1.6 ng/ml). The NE challenges resulted in small but significant reductions in tracer uptake (about 10% after atomoxetine, the cold stressor test and ketamine, no change by hypoglycaemia).

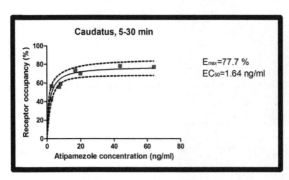

Inhibition of tracer uptake by different doses of atipamezole.

CONCLUSIONS: [^{11}C]ORM-13070 is a PET radioligand suitable for monitoring of α$_{2C}$-AR occupancy in the human brain. Its usefulness for monitoring NE release in the brain of healthy human subjects remains to be evaluated in larger future studies.

ACKNOWLEDGEMENTS: Supported by the Innovative Medicines Initiative Joint Undertaking (grant 115008) of EFPIA and the EU FP7 Programme.

DISCLOSURE: M. Scheinin and J.O. Rinne have contract research relationships with Orion Corporation (Espoo, Finland). M. Scheinin has received speaker's fees and research support from Orion Corporation.

Drug Abuse and Addiction

David Goldstein and Lee Eiden

It is generally accepted that the reinforcing behavior associated with drugs of abuse originates in central dopaminergic neurons. In a plenary lecture, Volkow described how over the past couple of decades the concept has emerged that addiction is actually independent of pleasure and that drug seeking can be triggered outside of conscious awareness. Addictive drugs first "hijack" the central process of dopamine- (DA)-dependent, reward-based learning but then go on to perturb the orchestrated balance of other neuronal circuits, including those necessary for insight, interoception, motivation, decision-making, and social behaviors. This functional overlap results in addicted individuals making poor choices despite awareness of the negative consequences; it explains why previously rewarding life situations and the threat of judicial punishment are inadequate to deter drug-taking; and why catecholamine modulating compounds are attractive targets for the development of addiction medications.

In a second plenary lecture also linked to Theme G, Wightman reviewed the history of electroanalytical chemistry for detection of catecholamines and in particular the use of microelectrodes for *in situ* sensing of catecholamines. The measurements offer the advantage of tracking catecholamine release, uptake, and signaling in subsecond time scales and can probe exocytotic events emanating from single cells. They also can be used in the brain of awake animals to follow catecholamine

Catecholamine Research in the 21st Century.

163

signaling during behavior. This approach has been used to confirm DA release in response to cues that predict reward as well as to examine norepinephrine signaling in aversive situations.

Bonci introduced the use of optogenetics to elucidate mechanisms of drugs of abuse and addiction. The mesolimbic and mesocortical systems play a role in motivation and reward-related behaviors and substance abuse. Combined optogenetics, electrophysiology, and behavioral procedures may provide clues for how these systems operate and potential treatments for craving and compulsive drug taking. Witten described a general approach for achieving specific optogenetic control of genetically defined cell types in rats. Multiple tyrosine hydroxylase (TH)::Cre and choline acetyl-transferase (ChAT)::Cre rat lines were generated that exhibited high cell-type specificity of opsin expression when injected with Cre-dependent ChR2-YFP-expressing virus. Optical stimulation of ChR2 in DA neurons in the ventral tegmental area of TH::Cre rats was sufficient to support vigorous intracranial self-stimulation. Zhang et al. reported evidence for dual neuronal signaling by DA neurons in the VTA, by release of DA, glutamate or co-release of DA and glutamate.

Cadet et al. reported on methamphetamine (METH) addiction mechanisms. METH addiction is a chronic relapsing disorder related to altered synaptic plasticity in DA and glutamatergic (Glu) striatal projection areas. The dorsal striatum integrates signals that mediate long-term changes in goal-directed behaviors. Epigenetic modifications secondary to drug abuse are important emerging mechanisms for explaining the molecular bases of addiction. The results provided evidence for a role of histone hypoacetylation in METH-induced decreased expression of GluA and GluN1 receptor subtypes. Casey et al. found attenuated amphetamine (AMPH)-induced DA release, based on ^{11}C-raclopride response to acute AMPH, in people at high risk for substance dependence. Cocaine (COC) is well known to increase extracellular fluid DA levels, and Fotros et al. used ^{18}F-Fallypride PET scanning to detect cue-induced DA release in the amygdala and hippocampus of COC-dependent users, while Nader et al. noted that long-term COC exposure affects D3 receptor function.

AMPH is a substrate for the DA cell membrane transporter (DAT), promoting the reversal of DA transport and thereby resulting in DA efflux via DAT. This efflux leads to increased extracellular DA levels, an event of importance for psychomotor stimulant properties. Hamilton et al. hypothesized that upon phosphorylation, the DAT N-terminus uncouples from PIP2 to become unconstrained, supporting DA efflux. By impairing DA efflux but not uptake, they determined the contribution of DA efflux in AMPH behaviors in *Drosophila*. Amara et al. noted that amphetamines, unlike COC and other non-transported blockers, have the ability to stimulate internalization of DAT from the cell surface. They found that cytoplasmic AMPH transiently activates Rho in a manner correlated with DAT internalization. Pretreatment of mice with agents that selectively increase cAMP in dopaminergic neurons reduces AMPH-evoked behaviors without altering those induced by COC. The activation of cytoplasmic signaling cascades contributes to behavioral effects of acute AMPH exposure and may explain some of the unique neurobiological consequences associated with AMPH use and abuse. Gether studied how rewarding effects of psychostimulants such as COC and AMPH result from molecular interactions with the DAT. Recent work has demonstrated the importance of C-terminal protein-protein interactions for DAT trafficking. The results indicate a critical role of Ca^{2+}/calmodulin dependent kinase IIα in AMPH action and an indispensable role of PDZ domain binding interactions in striatal distribution of the DAT, since DAT knock-in mice with disrupted PDZ-binding motifs had a dramatic loss of DAT expression in the striatum, causing hyperlocomotion upon exposure to a novel environment and loss of response to AMPH. Guptaroy et al. examined growth associated protein-43 (GAP-43) in AMPH actions—specifically the role of AMPH-stimulated GAP-43 phosphorylation in protein kinase C-beta-mediated AMPH regulation of DAT function, and found evidence that the phosphorylation enhances reverse transport of DA at the DAT in response to AMPH.

Martinez et al. noted commonalities and differences across addictions in central neuroimaging studies. Across different types of addiction, including COC, alcohol, and heroin dependence, addiction is associated with a decrease in D2/D3 receptors and a decrease in pre-synaptic DA release; however, imaging studies also show that in COC abuse blunted DA transmission predicts COC seeking. Low D2/D3 receptor binding and low DA release are associated with the choice to self-administer COC over alternative reinforcers, which can be viewed as a failure to shift between competing rewards. In contrast, a similar study in heroin dependent individuals did not show an association between these measures of DA transmission and heroin self-administration. Narendran cited preclinical literature about sensitization, an enhanced DA response in the nucleus accumbens that occurs after repeated exposure to psychostimulant drugs; neuroimaging studies of humans suggest blunted DA responses, the opposite of sensitization, in COC-dependent subjects. Observations by London et al. on links between striatal DA system markers, cognitive performance, and neural activity related to decision-making support an important role of D2-like receptors in emotional states and cognitive processes related to addiction, so that enhancement of signaling through these receptors may be a fruitful therapeutic objective.

Regarding treatment for addiction, Newman et al. reported on drug design targeting the D3 receptor (a member of the D2-like receptor family) with a focus on COC and METH. A recent focus has been on the D3 receptor, because of its restricted distribution to limbic regions of the brain. Nader et al. reported that long-term COC and MATH exposure affects D3 receptor function and on development of D3 partial agonists. In contrast with quinpirole, a D3 agonist that induces yawning in monkeys regardless of COC history, a D3 partial agonist elicited yawns only in monkeys with a COC history and decreased drug self-administration in a drug/food choice experiment. D3 partial agonists seem to be differentially effective *in vivo*, depending on the drug history of the individual and on the conditions of drug availability.

Selective D2, D3, and D4 antagonists and partial agonists have been discovered. Acri described development of a selective D4 ligand. LeFoll et al. studied D3 and D4 receptor antagonists for treatment of nicotine addiction. "Functional selectivity" (also known as biased signaling) refers to the process by which G-protein-coupled receptor ligands differentially modulate canonical and non-canonical signal transduction pathways. Sassano et al. reported on D2 functionally selective, β-arrestin-biased and Gi-biased ligands. Robbins discussed the hypothesis that cognitive dysfunction associated with chronic drug abuse derives from D2 receptor down-regulation associated with orbitofrontal hypometabolism, so that treatment with a D2/D3 agonist may under some conditions improve fronto-striatal functioning and associated executive function.

Addiction: Story of an Unbalanced Brain

Nora Volkow

National Institute on Drug Abuse (NIDA), Washington, D.C., USA

The past couple of decades of neuroscience research have transformed our understanding of why some people gravitate towards and eventually become addicted to psychoactive substances. The emerging picture suggests that addiction is actually independent of pleasure, and that drug seeking can be triggered outside of conscious awareness. We now know that addictive drugs first hijack the central process of dopamine dependent reward-based learning but go on to perturb the carefully orchestrated balance among many other neuronal circuits, including those necessary for insight, interoception, motivation, decision making and social behaviors. This functional overlap results in addicted individuals making poor choices despite awareness of the negative consequences; it explains why previously rewarding life situations and the threat of judicial punishment are ill suited to deter drug taking; why catecholamine modulating compounds are attractive targets for the development of addiction medications; and why a medical rather than a criminal approach would be far more effective in curtailing addiction.

Essential Catecholamine Influences on Dorsolateral Prefrontal Cortical Cognitive Function: Molecular Signaling Pathways

Amy Arnsten

Yale University School of Medicine, USA

Catecholamines are essential to the higher cognitive functions of the dorsolateral prefrontal cortex (dlPFC; Robbins and Arnsten, Annu Rev Neurosci. 32:267–87, 2009), a region that subserves our highest order cognitive abilities. The groundbreaking discovery by Brozoski et al. showed that depletion of catecholamines from the dlPFC was as devastating as removing the cortex itself (Science 205:929–32, 1979). Since then, we have learned how norepinephine (NE) and dopamine (DA) influence these highly evolved cortical circuits at the level of ion channels. These data show that dlPFC neurons are modulated in a fundamentally different manner than sensory-motor and subcortical circuits, and that genetic or environmental insults to this process are associated with cognitive disorders. Our research combines a variety of techniques to reveal modulatory influences on PFC circuits: multiple label immunoelectron microscopy to localize protein interactions, single unit recordings from the dlPFC of cognitively-engaged monkeys coupled with iontophoretic application of drug to assess physiological interactions, and behavioral assays following systemic or intra-PFC infusion of drug in monkeys and rodents to assess effects on working memory performance. The work is based on the anatomical and physiological studies of Goldman-Rakic (Neuron 14:477−85, 1995), which revealed recurrent layer III pyramidal cell microcircuits that maintain persistent firing through synapses on spines. We have recently shown that these synaptic connections depend on NMDA receptors, including those with NR2B subunits (Wang et al, in submission). Neuronal persistent firing is tuned by lateral inhibition from GABAergic interneurons (ibid). These layer III microcircuits are greatly afflicted in schizophrenia (reviewed in Arnsten, Int J Dev Neurosci. 29:215−23, 2011). We have found that layer III spines are long and thin and greatly enriched in Ca-cAMP signaling proteins, including Ca- and cAMP-modulated potassium channels that gate network inputs (Arnsten et al., TICS 14:365−75, 2010; Paspalas et al. Cereb Cortex 2012 Jun 12. [Epub ahead of print]). NE strengthens network connections and improves PFC cognitive functions through a2A-AR inhibition of cAMP-HCN and KCNQ channel signaling (Wang et al. Cell 129:397−410, 2007; Nature 476:210−3, 2011). This work has led to the use of the a2A-AR agonist, guanfacine, for the treatment of PFC disorders in humans (Arnsten, Expert Rev Neurother. 10:1595−605, 2010). Conversely, DA D1 has an inverted-U dose response, gating out "noise" at optimal doses, but suppressing all firing at higher doses, as occurs with stress (Vijayraghavan et al. Nat Neurosci. 10:376−84, 2007; Arnsten, Nat Rev Neurosci. 10:410−22, 2009). High, "stressful" levels of NE also suppress firing and impair working memory via a1-Ca-PKC signaling (Birnbaum et al. Science 306:882−4, 2004). Layer III synapses are also enriched in proteins that regulate Ca-cAMP signaling (e.g. RGS4, DISC1-PDE4A). Importantly, many of these proteins are genetically altered in schizophrenia, and may contribute to weakened dlPFC function and symptoms of thought disorder (Arnsten, Int J Dev Neurosci. 29:215−23, 2011). Alterations in DA D2 modulation of feedback from layer V Response cells may also contribute to symptoms such as hallucinations (Wang et al. Science 303:853−6, 2004). Understanding the unique modulation of these higher cortical circuits will be essential for linking genetic insults to symptoms, and devising rational treatments for cognitive disorders.

Neural Mechanisms Underlying the Cognition-Enhancing Actions of Pyschostimulants: Relevance to ADHD

Craig W. Berridge, David M. Devilbiss, Robert C. Spencer and Brooke E. Schmeichel

Psychology Department, University of Wisconsin, Madison, WI 53706, USA

Psychostimulants are highly effective and widely used to treat attention deficit hyperactivity disorder (ADHD). These drugs improve a variety of prefrontal cortex (PFC)-dependent processes in humans with and without ADHD. Despite their longstanding clinical use, the neural mechanisms responsible for their therapeutic actions have only recently begun to be examined. We have shown that, in rats, doses of methylphenidate (MPH; Ritalin) that result in clinically-relevant blood concentrations improve performance in a PFC-dependent task of working memory similar to that seen in humans. At these clinically-relevant doses, MPH elevates extracellular catecholamine levels and enhances neuronal signal processing preferentially within the PFC (Berridge et al., 2006; Devilbiss & Berridge, 2008, Biolog. Psychiat. 60:1111-1120; 64:626-635) while preserving sensory-driven phasic responsiveness of locus coeruleus noradrenergic neurons (Devilbiss & Berridge, 2008, JPET 319:1327-1335). The regionally selective actions of clinically-relevant doses of psychostimulants contrast with the widespread effects on catecholamine signaling observed at higher doses. The preferential elevation of both norepinephrine (NE) and dopamine (DA) within the PFC by low-dose psychostimulants likely reflects a prominent role of the NE transporter in the clearance of extracellular DA within this region. Consistent with this, we recently observed that direct perfusion of low concentrations of MPH through a microdialysis probe similarly elicits larger increases in extracellular NE/DA within the PFC than outside the PFC. Moreover, we recently demonstrated that a selective DA transporter inhibitor, which lacks abuse potential associated with psychostimulants, elevates extracellular levels of both DA and NE within the PFC at a dose that improves working memory performance (Schmeichel, Berridge, in press). Collectively, these observations suggest the cognition-enhancing actions of psychostimulants involve, at least in part, direct action within the PFC. Consistent with this, we recently observed that microinfusion of MPH into the dorsomedial, but not ventromedial, PFC improves working memory performance (Spencer et al., 2011, Biolog. Psychiat. e-pub ahead of print: doi:10.1016/j.biopsych.2011.07.022). Recently completed studies in our lab demonstrate that the cognition-enhancing actions of intra-PFC infused MPH are dependent on the activation of PFC a2and D1 receptors. Finally, we have observed that although the dorsomedial and ventromedial striatum are necessary for performance of this task, MPH action within these regions is insufficient to improve working memory performance (Spencer et al., ibid). Collectively, these observations indicate that the cognition-enhancing actions of psychostimulants are, at least in part, dependent on increased signaling at a2and D1 receptors within the PFC. Available evidence suggests that there are differential dose-dependent actions of MPH on separable cognitive/behavioral processes (Berridge et al, 2011, Biolog. Psychiat, 71:467-473). In particular, MPH improves PFC-dependent working memory performance and response inhibition in a narrow inverted-U-shaped dose-dependent manner. In contrast, overt behavior and select other PFC-dependent processes, such as sustained attention and attentional set shifting, display righted-shifted inverted-U sensitivity to MPH. Interestingly, these higher dose actions are linked to the activation of lower affinity a1-receptors (relative to a2-receptors). These latter observations suggest the possibility that certain cognitive/behavioral processes may be impaired at higher doses that maximally control behavioral activity and focus attention. These studies provide new insight underlying the cognition-enhancing actions of psychostimulants that may be particularly relevant for the development of novel pharmacological treatments for ADHD and other conditions associated with PFC dysregulation.

Frontostriatal Systems and Differential Roles of Dopamine D1- and D2-Like Receptors in Impulsivity and Response Inhibition

Edythe D. London[1,2,3], **Chelsea L. Robertson**[2,3], **Kenji Ishibashi**[1,2], **Mark A. Mandelkern**[2,4], **Dara Ghahremani**[1,2], **Eliza Congdon**[1,5], **Fred Sabb**[1,5], **Andy C. Dean**[1], **Amira K. Brown**[1] and **Tyrone Cannon**[1,5]

[1]*UCLA Semel Institute of Neuroscience and Human Behavior, USA;* [2]*West Los Angeles Veterans Administration of Greater Los Angeles, USA;* [3]*UCLA Department of Molecular and Medical Pharmacology, USA;* [4]*UC Irvine Department of Physics, USA;* [5]*UCLA Department of Psychology, USA*

Patients with addictions and other neuropsychiatric disorders that are associated with deficits in self-control exhibit dopaminergic dysfunction, and studies in humans as well as animals have indicated that cortical dopamine D1-like receptors (D1Rs) (1) and striatal D2-like receptors (D2Rs) (2,3) influence self-control and conversely impulsivity. Response inhibition, a measure of self-control, requires striatal dopaminergic neurotransmission; and a recent animal study showed opposing roles for striatal D1Rs and D2Rs in response inhibition (4). The relative contributions of these systems to self-control (and impulsivity), however, are not completely understood. We addressed this question using positron emission tomography (PET) with [11C]NNC112 and [18F]fallypride to measure availability of D1Rs and D2Rs, respectively, in healthy subjects who gave self-reports on the Dickman Impulsivity Inventory, including functional and dysfunctional impulsivity scales, and performed the Stop-signal task, a test of response-inhibition capacity. This ongoing study has included 16 participants (9 men, 7 women), 29.6 ± 8.8 (mean $\pm$ SD) years old. They underwent PET scans with the two tracers, and MRI scans for co-registration of PET data. Striatal volumes of interest (VOIs) were defined on individual MRI scans and transferred to the co-registered PET images. The PMOD software package was used to calculate binding potential (BPND) using the simplified reference tissue model with the cerebellum as the reference region. In addition, BPND maps were analyzed using SPM8. Scores on the Dickman scales and stop-signal reaction time (SSRT), indexing response inhibition capacity on the Stop-signal task, were determined, and partial correlations (controlling for age) were performed to assess their relationships to BPND values. Whereas functional impulsivity was not related to either striatal D1R BPND or to D2R BPND, dysfunctional impulsivity was negatively correlated with D2R BPND in the caudate nucleus ($r = 0.64$, $p = 0.02$), putamen ($r = 0.65$, $p = 0.02$), and accumbens ($r = 0.61$, $p = 0.03$). Voxel-wise analysis confirmed the negative correlation involving striatal D2R BPND ($p < 0.01$, uncorrected). VOI analysis showed no significant correlation of dysfunctional impulsivity with striatal D1R BPND, but voxel-wise analysis showed a positive correlation with D1R BPND in the medial orbito-frontal cortex ($p < 0.01$, uncorrected). VOI analyses showed no significant relationship between striatal D1R BPND and SSRT, and a significant correlation between caudate D2R BPND with SSRT ($r = -0.69$, $p = 0.019$). Voxel-wise regression analysis confirmed a negative correlation of SSRT with D2R BPND in the dorsal striatum, at a liberal threshold ($p < 0.05$, uncorrected). A similar voxel-wise regression analysis of D1R BPND showed a positive relationship with SSRT in a cortical area extending from the pre-supplementary motor area to the dorsal anterior cingulate, at the same liberal threshold ($p < 0.05$, uncorrected). These observations extend findings that D1Rs in the medial orbitofrontal cortex and D2Rs in the striatum play important roles in impulsivity (1–3), particularly dysfunctional impulsivity, and suggest that D1Rs and D2Rs have differential actions. They support the view that striatal D2Rs contribute to response inhibition, and provide preliminary evidence that D1Rs in the cortex may play a role in the neurocircuitry that underlies response inhibition.

References

1. Loos M, Pattij T, Janssen MC, et al. (2010): Cereb. Cortex 20(5); 1064–1070.
2. Lee B, London ED, Poldrack RA, et al. (2009): J. Neurosci. 29(47):14734–40.
3. Dalley JW, Fryer TD, Brichard L, et al. (2007): Science 315(5816):1267–1270.
4. Eagle DM, Wong JC, Allan ME, et al. (2011): J. Neurosci. 31(20):7349–7356.

Catecholamine Strategies for Remediating Fronto-Executive Deficits in Stimulant Addiction and Related Disorders

TW Robbins

Behavioural and Clinical Neuroscience Institute and Dept. of Expt. Psychology, University of Cambridge, UK

Stimulant drug addiction is associated with fronto-executive dysfunction, but it is unclear to what extent this is the consequence of chronic drug abuse or is predisposing to it. I will describe strategies directed towards disentangling these causal influences in both rodents and humans. One hypothesis of cognitive function associated with chronic drug abuse derives from observations of dopamine D2 down-regulation associated with orbitofrontal hypometabolism, suggesting that treatment with a D2/3 agonist under some conditions may improve fronto-striatal functioning and associated executive function. Treatment with the D2/3 agonist pramipexole (0.5 mg) reduced drug Stroop interference and associated prefrontal cortex activation in chronic cocaine abusers who scored highly on the OCDUS questionnaire of compulsive drug use (Ersche et al 2010). In a parallel study, the D2/3 agonist also reduced perseveration and associated hypoactivation of the anterior caudate nucleus in cocaine abusers in a test of reversal learning (Ersche et al 2011a). In a third study, we examined effects of pramipexole on other tests of neurocognitive function from the CANTAB battery. Stimulant users performed worse than healthy volunteers on the cognitive tests. The variation in whole blood peripheral dopamine D3 receptor mRNA expression explained over one quarter of the variation in response to pramipexole on the spatial working memory test across all participants, suggesting that peripheral biomarkers could also be used to predict responses to the agonist (Ersche et al 2011b). An alternative strategy will utilise the relatively selective noradrenaline reuptake blocker atomoxetine, based on the observation that this drug normalizes impulsivity in rats that is associated with compulsive cocaine seeking (Fernando et al 2012). Additionally, atomoxetine reduces both cocaine and heroin drug-seeking in rats at doses that do not grossly affect behavior (Economidou et al 2011). The anti-impulsive effects of atomoxetine have also been linked to both the shell region of the nucleus accumbens and also certain regions of the prefrontal cortex (in healthy human volunteers, as well as rats). Finally, a study of nearly 2000 adolescents of stop-signal performance in an fMRI setting shows some overlap in activity of prefrontal circuits engaged in executive control by adolescents using drugs (including cocaine) and those associated with a polymorphism of the noradrenaline transporter (Whelan et al 2012). These studies illustrate possible contributions of both dopamine and noradrenaline to aspects of fronto-executive control that may be relevant to the remediation of behavioral deficits in chronic drug abusers, and illustrate strategies for resolving the heterogeneity of response to candidate therapeutic drugs for drug abuse based on both genetic and neuroimaging approaches.

References

Ersche,K.D., et al.(2010) Archives of General Psychiatry, 67, 632−644.

Ersche,K.D., et al.(2011a) Biological Psychiatry, 70, 754−762.

Ersche,K.D., et al.(2011b). Psychopharmacology, 214, 779−789.

Fernando,A.B., et al.(2012) Psychopharmacology, 219, 341−352.

Economidou D. et al.(2011) Biol. Psychiat. 69, 266−274.

Whelan, R. et al.(2012) Nature Neuroscience 15, 920−U153.

Ultrastructural and Optogenetic Evidence for Dual Neuronal Signaling by Dopamine Neurons of the Ventral Tegmental Area (VTA)

Shiliang Zhang[1], Xueping Li[1], Huiling Wang[1], Jia Qi[1], Jonathan P Britt[2], Antonello Bonci[2] and Marisela Morales[1]

[1]*National Institute on Drug Abuse, Inrtamural Research Program, Neuronal Networks Section, 251 Bayview Blvd., Suite 200, Baltimore, Maryland , United States;* [2]*National Institute on Drug Abuse, Intramural Research Program, Synaptic Plasticity Section, 251 Bayview Blvd., Suite 200, Baltimore, Maryland, United States*

BACKGROUND: Dopamine neurons in the midbrain were discovered more than 50 years ago. These neurons express tyrosine hydroxylase (TH) and play a role in movement, motivation, learning and reward. We previously found that in addition to the classical dopamine neurons, specific regions of the midbrain (such as the VTA) contain neurons expressing vesicular glutamate transporter 2 (VGluT2) mRNA. VGluT2 selectively transports glutamate into synaptic vesicles, thus, neurons containing VGluT2 use glutamate as neuronal signaling neurotransmitter. There are 2 classes of VGluT2 neurons: the VGluT2-only neurons (containing VGluT2 mRNA without TH) and dual TH-VGluT2 neurons (co-expressing TH and VGluT2 mRNA). Both types of VGluT2 neurons target the medial prefrontal cortex (mPFC) and the nucleus accumbens (nAcc), brain structures involved in drug abuse. These findings indicate that neurons of the VTA innervating the mPFC or the nAcc have the capability to (a) release dopamine, (b) release glutamate or (c) co-release dopamine and glutamate. Co-release of dopamine and glutamate could constitute a novel mechanism of neuronal signaling.

OBJECTIVES: To investigate whether neurons co-expressing TH and VGluT2 mRNA mediate synaptic co-release of dopamine and glutamate.

METHODS AND RESULTS: In a first set of studies, we performed in vivo tagging of VTA neurons in the rat by either region specific injections of tract tracing molecules or viral vectors. The specific cellular tagging together with immunolabeling methods allowed us to identify at the electron microscopic level the axon terminals derived from VTA neurons, and determine their morphological and chemical characteristics. Findings from these approaches showed expression of TH in axon terminals lacking VGluT2, and expression of VGluT2 in axon terminals lacking TH. Next, to determine whether lack of detection of co-existence of TH and VGluT2 proteins within the same axon terminal was due to lack of translation of VGluT2 mRNA in these TH-VGluT2 neurons or due to subcellular segregation of TH and VGluT2, we next induced expression of the light activated opsin channelrhodopsin-2 tethered to m-Cherry (ChR2-mCherry) in TH or VGluT2 neurons following viral infection into the VTA of TH::Cre mouse or VGluT2::Cre mouse. Immuno ultrastructural analysis of these brains showed that all axon terminals from VTA tagged neurons making asymmetric synapses contained VGluT2 protein but lacked TH, and that those making symmetric synapses contained TH but lacked VGluT2. These findings indicate that indeed TH-VGluT2 neurons have the capability to synthesize VGluT2 protein, and that this protein gets incorporated into vesicles located in axon terminals for the accumulation and synaptic release of glutamate. These findings also indicate that TH axon terminals lack VGluT2, however, TH was observed in the contiguous axons of axon terminals containing VGluT2. Finally, by combination of ex vivo electrophysiology and optogenetics (light induced activation of fibers from VTA neurons expressing ChR2-mCherry in TH or VGluT2 neurons), we found that the above identified axon terminals from VTA VGluT2 neurons innervating the mPFC or nAcc evoked excitatory postsynaptic currents.

CONCLUSIONS: We provide for the first time ultrastructural, optogenetic and electrophysiological evidence indicating that a brain area involved in addictive behaviors has a unique set of neurons with the unanticipated capability to co-release two different signaling molecules from two distinct subcellular compartments: glutamate from axon terminals, and dopamine mostly from axons. Further studies are necessary to determine the role of this novel dual signaling mechanism in brain function.

Seeking reward: Upstream modulation of VTA dopamine neurons by orexin/hypocretin inputs

Gary Aston-Jones, Ph.D., Steve Mahler, Ph.D., Greg Sartor, Ph.D., Rachel Smith, Ph.D. and David Moorman, Ph.D.

Medical University of South Carolina, Charleston, SC

The orexins (also known as hypocretins) are neuropeptide transmitters in hypothalamic neurons. Many studies have implicated orexin neurons in arousal, but there is also pronounced orexin innervation in reward-associated brain regions including ventral tegmental area (VTA). We found that Fos expression in lateral hypothalamus (LH) (but not more medial) orexin neurons was strongly correlated with preferences for stimuli associated with morphine, cocaine or food reward (Harris et al Nature

437: 556, 2005). Chemical activation of LH orexin neurons reinstated an extinguished conditioned place preference (CPP) for morphine. This reinstatement was blocked by pretreatment with the OxR1 antagonist, SB334867. Moreover, microinjection of orexin directly into VTA also reinstated an extinguished morphine CPP. SB334867 also blocked stimulus-induced (but not cocaine-induced) reinstatement of extinguished cocaine-seeking in a self-administration paradigm (Smith et al Eur. J. Neurosci. 30: 493, 2009). This effect was not seen for a selective OxR2 antagonist. Borgland et al. (Neuron 49: 589, 2006) showed that orexin potentiates glutamate responses of VTA dopamine (DA) neurons. Our neurophysiology studies revealed that orexin potentiates responses of VTA dopamine neurons to medial prefrontal cortex stimulation, a possible source of stimulus reward information conveyed to VTA. Also, we discovered that disruption of the LH orexin projection to VTA prevented learning a morphine CPP (Harris et al Behav. Brain Res. 183: 43, 2007), consistent with previous findings that orexin is critical for drug-associated LTP in VTA DA neurons (Borgland et al 2006). Recently, we also found that inputs to LH orexin neurons from lateral septum (LS) are necessary for the Fos activation of orexin neurons by expression of cocaine CPP, and that interruption of this circuit prevents cocaine preference in this paradigm (Sartor & Aston-Jones, J. Neurosci. 32: 4623, 2012). These data reveal a novel role for an LS- LH orexin—VTA circuit in stimulus-reward relationships that are important in reward-seeking, drug relapse and addiction.

Striatal Dopamine, Self-control and Decision-Making: Identifying Therapeutic Targets for Addiction through Brain Imaging

Edythe D. London[1,2], Mark A. Mandelkern[2,3], Dara Ghahremani[4], Milky Kohno[4], Angelica Morales[4], Chelsea L. Robertson[2,5] and Kenji Ishibashi[1,2]

[1]*UCLA Semel Institute of Neuroscience and Human Behavior, USA;* [2]*West Los Angeles Veterans Administration of Greater Los Angeles, USA;* [3]*UC Irvine Department of Physics, USA;* [4]*UCLA Interdepartmental Neuroscience Training Program, USA;* [5]*UCLA Department of Molecular and Medical Pharmacology, USA*

A role of striatal dopamine in addictive disorders has been well established through studies of animal models. The evidence included demonstrations that self-administration of drugs of abuse increased extracellular dopamine concentrations and was attenuated by administration of dopamine receptor antagonists and selective lesions. With the advent of noninvasive molecular neuroimaging techniques, such as positron emission tomography (PET), human studies have extended this research. Participants who meet criteria for a variety of addictive disorders, including stimulant dependence and pathological obesity, all show lower striatal D2-like dopamine receptor availability when compared to healthy control participants. These findings led to the hypothesis that dysfunction in mesocorticolimbic dopamine circuits can promote a "Reward Deficiency Syndrome", characterized by anhedonia and aberrant behaviors that are linked to addiction. This view was supported by observations that drug-induced euphoria accompanies striatal dopamine release, and that when one reward is no longer available, another reward is substituted. For example, individuals in early abstinence from chronic meth-amphetamine (METH) abuse apparently substitute food for their drug of choice. They exhibit unusually high caloric intake, which is inversely correlated with striatal D2-like dopamine receptor availability. Beyond its role in reward processing, dopamine has been a focus in studies of self-control, which is impaired in addicts who cannot stop or control their drug use. In this regard, METH-dependent research participants rate themselves as more impulsive than their non-drug-abusing counterparts on the Barratt Impulsiveness Scale, and these self-reports are inversely correlated with striatal D2-like receptor availability. On average, METH-dependent individuals also perform worse than control subjects on tests of response inhibition, cognitive flexibility and other executive functions. Notably, performance deficits in METH users have been observed on the Stop-signal task, in which stopping ability and inhibition-related fMRI activation in frontostriatal neural circuitry are correlated with D2-like receptor availability in the striatum, consistent with findings of animal studies relating dopamine receptors and response inhibition performance. METH users also have deficits in cognitive flexibility, as indicated by performance on a reversal−learning task. In vervet monkeys, D2- but not D1-like receptor antagonists impair post-reversal performance without influencing the acquisition of new associations, suggesting a selective role for D2-like receptors in cognitive flexibility. Moreover, METH produces decreases in striatal D2-like receptor availability as well as reversal-learning deficits in monkeys. Another consideration with respect to addiction is faulty decision-making, which can reflect a propensity for risk-taking. Our research team recently completed a study of brain activation associated with risk-taking in 60 healthy human subjects using an fMRI-compatible version of the Balloon Analogue Risk Task and PET assessment of D2-like receptor availability in the striatum. We observed that striatal D2-like receptor availability was negatively correlated with prefrontal sensitivity of activation to the levels of risk and reward when a participant decided to take risk. Earnings on the task were related positively to risk/reward sensitivity of prefrontal activation and negatively to striatal D2-like receptor availability, suggesting an interactive system that supports adaptive risk-taking. These observations on the links between striatal dopamine system markers, cognitive performance and neural activity related to decision-making support an important role of D2-like dopamine receptors in feeling states and cognitive processes related to addiction. Overall, they support the view that enhancement of signaling through these receptors may be a fruitful therapeutic objective.

Imaging Neurochemical Transmission in Substance Abuse: Commonalities and Differences Across Addictions

Diana Martinez[1], Alexander Grassetti[1], Yiyun Huang[2], Mark Slifstein[1], Lawrence Kegeles[1], Richard Carson[2], Kenneth Carpenter[1], Edward Nunes[1] and Richard Foltin[1]

[1]*Columbia University/New York State Psychiatric Institute, USA;* [2]*Yale University School of Medicine, USA*

The involvement of dopamine in addiction has its origins in studies investigating reward and reinforced behavior in preclinical studies. Much of this research has been explored in the human brain using Positron Emission Tomography (PET) imaging of striatal dopamine transmission. These studies show that addiction is associated with a decrease in dopamine D2/3 receptors and a decrease in presynaptic dopamine release, and that this decrease occurs across different types of addiction, including cocaine, alcohol, and heroin dependence. However, these imaging studies also show that, in cocaine abuse, blunted dopamine transmission is predictive of cocaine seeking behavior. Low D2/3 receptor binding and low dopamine release are associated with the choice to self-administer cocaine over alternative reinforcers (such as money), which can be viewed as a failure to shift between competing rewards. However, a similar study in heroin dependent individuals did not show a similar association between these measures of dopamine transmission and heroin self-administration. A recent study in cocaine abusers also showed that this disorder is associated with a decrease in metabotropic glutamate receptors, but that this also did not correlate with cocaine self-administration, suggesting that this behavior may be associated with alterations in dopamine, but not glutamate, transmission. Overall, this data suggests that increasing dopamine signaling at the striatal D2/3 receptor s may serve as an appropriate treatment approach to cocaine abuse. However, the best mechanism for achieving this remains unclear. Pre-clinical studies have shown that cocaine exposure is not associated with injury of the dopamine neurons, suggesting that they remain intact in the human brain. In addition, numerous clinical trails have not shown that non-specific dopamine agonists have an impact on treatment. Thus, studies investigating the neurochemical systems that regulate dopamine release are needed, as are studies that investigate ways to increase post-synaptic dopamine signaling.

Dopamine Transmission in Cocaine Addiction: Linking the Animal and Human Studies

Raj Narendran
UPMC, USA

Much effort has been devoted in the preclinical addiction literature to understanding the phenomenon of sensitization, an enhanced dopaminergic response in the nucleus accumbens that occurs after repeated exposure to psychostimulant drugs.Imaging studies with PET and SPECT using a stimulant challenge to induce dopamine release provide a unique opportunity to probe dopamine transmission in cocaine dependent human subjects. In contrast to the basic science literature that predicted sensitization, results of the imaging studies suggest blunted dopamine response, or the opposite of sensitization, in human cocaine dependent subjects. This presentation will review the methodological differences between the preclinical and clinical PET studies that have investigated dopamine sensitization in cocaine dependence. In addition, it will discuss more recent imaging studies that were done to understand the discrepancy between the human and animal literature. Finally, the presentation will discuss how these imaging results can be used to guide drug development efforts in cocaine addiction.

Growth Associated Protein-43, Amphetamine and Dopamine Transporter Activity

Bipasha Guptaroy, Aalisha Desai and Margaret E. Gnegy
University of Michigan, USA

BACKGROUND: Psychostimulants such as amphetamine (AMPH) produce their rewarding and euphoric effects by acting at the presynaptic dopamine transporter (DAT). AMPH competitively inhibits DA reuptake and elicits outward transport of DA (efflux), increasing synaptic DA content, which is significant for its psychostimulant action. Protein Kinase C (PKC) specifically PKCβ, which associates with DAT, mediates in part the action of AMPH on DA efflux. A selective PKCβ inhibitor or PKCβ gene deletion in mice reduces AMPH-stimulated DA efflux. A prominent presynaptic substrate of PKCβ that is phosphorylated at its PKC phosphorylation site (ser41) by both acute and repeated AMPH stimulation is growth-associated protein 43 (GAP-43). This underscores the potential relevance of GAP-43 in AMPH action and suggests that GAP-43 phosphorylation by PKCβ could mediate AMPH action on DAT. Phosphorylation of GAP-43 at ser41 is essential for vesicular trafficking and cytoskeletal associations which are likely also involved in AMPH regulation of DAT function. Because GAP-43 is enriched in DA neurons and is a prominent substrate for PKCβ, it is a potential downstream effector for PKCβ in DAT reverse transport. Involvement of GAP-43 in AMPH action has further implications in plasticity induced by repeated AMPH exposure since phosphorylation of GAP-43 is implicated in synaptic plasticity.

OBJECTIVES: The objective is to determine the role of AMPH-stimulated GAP-43 phosphorylation in PKCβ-mediated AMPH regulation of DAT function.

METHODS AND RESULTS: AMPH-stimulated DA efflux was measured in transfected hDAT-HEK cells, by superfusion and in attached cells. Cells were treated with AMPH (10 μM) for 2 min and fractions collected and analyzed for DA content by HPLC with electrochemical detection, against an internal standard. PKCβ inhibitor LY379196 was used at 200 nM. Cell lysate proteins were separated by SDS-PAGE, transferred to nitrocellulose membranes and immunoblotted with anti-GAP-43 or anti phosphoser-41GAP-43 and appropriate HRP-conjugated secondary antibodies and detected by enhanced chemiluminisence. [3H]DA uptakeand [3H]WIN35,428 binding (as a measure of cell surface DAT) was measured in attached cells using varying concentrations or 1 μM DA [3H]DA (10 min) and 30 nM [3H]WIN32,428 (30 min) at room temperature. Surface DAT in rat synaptosomes and GAP-43 transfected cells was biotinylated with sulfo-NHS-SS biotin, lysed and biotinylated proteins were bound to avidin beads. Avidin-bound proteins were eluted and immunoprecipitated with anti-DAT and Protein A sepharose and subjected to immunoblotting with anti-GAP-43 and anti-DAT. The role of GAP-43 phosphorylation in AMPH-stimulated DA efflux was investigated in hDAT-HEK cells, lacking endogenous GAP-43. AMPH-stimulated DA efflux was enhanced in GAP-43 transfected cells compared to vector transfected cells with no corresponding increase in DA uptake or cell surface DAT. Conversely, suppressing endogenous GAP-43 expression with siRNA in hDAT-N2A cells, results in decreased AMPH-stimulated DA efflux. AMPH increased phosphorylation of GAP-43 at ser41in hDAT-HEK cells. The role of GAP-43 phosphorylation in AMPH-stimulated DA efflux was determined using phosphomimetic mutants of GAP-43. DA efflux was enhanced by GAP-43 S41D (mimicking phosphorylation) but not by GAP-43 S41A (non-phosphorylatable mutant). Treatment with PKCβ inhibitor, LY379196, prevents both AMPH-stimulated increase in GAP-43 phosphorylation at ser-41and enhanced AMPH-stimulated DA efflux by GAP-43 in hDAT-N2A cells. Additionally, GAP-43 associates with surface DAT.

CONCLUSIONS: Phosphorylation of GAP-43 by PKCβ enhances reverse transport of DA at DAT.

The Role of Ion Binding in Transporter Conformational Transition

Lei Shi

Weill Medical College of Cornell University, USA

The eukaryotic members of the presynaptic neurotransmitter:Na+ symporter (NSS) family are responsible for the termination of neurotransmission through Na+-driven uptake of various neurotransmitters, including catecholamines (dopamine and norepinephrine). The dopamine, norepinephrine, and serotonin transporters are the molecular targets for psychostimulant drugs and for many antidepressants.Despite the availability of several high-resolution crystal structures of a bacterial homologue of NSSs — the Leucine Transporter (LeuT), and extensive computational and functional analysis, many questions remain about the transport mechanism and the associated conformational dynamics. Although the exact molecular mechanism of the Na+-induced conformational dynamics was not known, we had proposed that Na+ binding is prior to substrate binding and leads to a conformational transition of the transporter to a more outward-open conformation to facilitate the entrance of the substrate[1−3]. It has also been shown that the transport of LeuT is pH dependent and the absence of Na1 has been predicted computationally to result in the protonation of Glu290 in the inward-open conformation[4]. By using a combination of micro-second atomistic molecular dynamics (MD) simulations with free energy computations, we revealed ion-controlled conformational dynamics of the protein in relation to binding affinity and selectivity of the Na+ binding sites. We show that in the course of MD simulations starting from the occluded state of LeuT with bound Na+, but in the absence of any substrate, the extracellular vestibule of LeuT spontaneously transitions to an outward-open conformation. Under these conditions, from MM/PBSA computations and a correlation analysis, we found Na+ binding affinity at the Na1 site is significantly reduced compared to that of the occluded state in the presence of the bound substrate, and is inversely correlated with the extent of completion of the transition to the outward-open conformation. Together with other simulation results and analysis, we conclude Na2 but not Na1 plays a critical role in establishing the outward-open conformation. However, the outward-open scale of the conformation can be modulated by the protonation state of the Na1-associated Glu290[5]. Taken together, we can now draw a relatively comprehensive mechanistic picture regarding the roles of Na+ and protons in the transport cycle.

References

1. Shi, L.; Quick, M.; Zhao, Y.; Weinstein, H.; Javitch, J. A. The mechanism of a neurotransmitter:sodium symporter--inward release of Na+ and substrate is triggered by substrate in a second binding site. Mol Cell 2008, 30, 667−77.

2. Quick, M.; Yano, H.; Goldberg, N. R.; Duan, L.; Beuming, T.; Shi, L.; Weinstein, H.; Javitch, J. A. State-dependent conformations of the translocation pathway in the tyrosine transporter Tyt1, a novel neurotransmitter:sodium symporter from Fusobacterium nucleatum. J Biol Chem 2006, 281, 26444–54.
3. Claxton, D. P.; Quick, M.; Shi, L.; de Carvalho, F. D.; Weinstein, H.; Javitch, J. A.; McHaourab, H. S. Ion/substrate-dependent conformational dynamics of a bacterial homolog of neurotransmitter:sodium symporters. Nat Struct Mol Biol 2010.
4. Zhao, Y.; Quick, M.; Shi, L.; Mehler, E. L.; Weinstein, H.; Javitch, J. A. Substrate-dependent proton antiport in neurotransmitter:sodium symporters. Nat Chem Biol 2010, 6, 109–16.
5. Zhao, C.; Stolzenberg, S.; Gracia, L.; Weinstein, H.; Noskov, S. Y.; Shi, L. Dynamic mechanisms of Na + binding prepare an outward-open state for substrate entry in LeuT. (submitted).

Evidence for the Contribution of D4 Receptors to Cocaine Self-Administration in Monkeys

Jane Acri

Chief, Medication Discovery and Toxicology Branch, DPMC, NIDA, USA

The effects of cocaine on behavior are primarily mediated by inhibition of monoamine reuptake, and the resulting stimulation of postsynaptic monoamine receptors. Dopamine receptors have been the focus of a great deal of study to determine their contributions to the behavioral effects of cocaine related to its abuse. However, the contribution of the D4 receptor has been challenging to study in part because of the absence of selective ligands. The NIDA Addiction Treatment Discovery Program has protocols in place for the blinded evaluation of compounds for efficacy in several models of substance use disorders. One of the most important models uses rhesus monkeys trained to self-administer cocaine in a rapid extinction protocol. In this paradigm, monkeys are trained to self-administer a training dose of cocaine, and to quickly extinguish when saline is substituted for cocaine. In addition, the protocol employs food-reinforced sessions before and after the availability of cocaine. Using this model, we are able to evaluate the effects of compounds across the entire dose-effect curve of cocaine, and importantly, to evaluate the non-specific effects on responding for food. This protocol has been used to evaluate a number of pharmacological mechanisms of actual candidate medications as well as research tool compounds. A brief review of the drug abuse-related effects of dopamine D4 receptor stimulation and blockade will be presented, along with results of NIDA evaluation of D4 ligands in the cocaine self-administration protocol described above.

Drug Design for Addiction — Molecular Determinants of Selectivity and Efficacy at the Dopamine D3 Receptor

**Amy Hauck Newman[1], Ashwini K. Banala[1], Prashant Donthamsetti[2], Jianjing Cao[1],
Mayako Michino[3], Robert R. Luedtke[5], Jonathan A. Javitch[2] and Lei Shi[3,4]**

[1]Molecular Targetsand Medication Discovery Branch, Medicinal Chemistry Section, National Institute on Drug
Abuse- Intramural Research Program, Baltimore, MD 21224; [2]Center for Molecular Recognition and Departments
of Psychiatry and Pharmacology, Columbia University College of Physicians and Surgeons, New York, New York,
USA; [3]Department of Physiology and Biophysics; [4]Institute for Computational Biomedicine, Weill Medical College
of Cornell University, New York, NY, USA; [5]Department of Pharmacology and Neuroscience, University of North
Texas Health Science Center, Fort Worth, Texas, USA

The dopamine D3 receptor (D3R), a member of the dopamine D2 receptor family, has been investigated as a potential target for medication development to treat substance use disorders (SUDs) with a particular focus on cocaine and methamphetamine. Currently, there are no approved medications to treat cocaine and methamphetamine addiction and thus developing pharmacotherapeutics to compliment existing behavioral strategies is a fundamental goal. The selective distribution of D3R in limbic regions of the brain, especially the nucleus accumbens has fortified interest in the D3R as a potential target for drug discovery. However, the high degree of sequence identity within the transmembrane (TM) segments of the D2-like receptors and the near-identity of the residues that form the orthosteric binding site (OBS) in these receptors have made it challenging to create subtype-selective agents that possess physicochemical properties suitable for in vivo characterization of their physiological roles. Nevertheless, significant progress has been made for D2-subfamily receptor-selective ligands, including the discovery of selective D2R, D3R and D4R antagonists and partial agonists. Compounds such as SB 277011A, NGB 2904, BP 897 and PG01037 have provided both critical tools for further characterization of D3R in addiction and pharmacophoric templates for the evolution of subsequent generations of D3R-selective agents. Structure-activity relationships (SAR) have been derived through extensive medicinal chemistry efforts, resulting in highly potent and D3R-selective agents with varying intrinsic activities. Classical drug design, based on SAR, has recently been supplemented with the ability to use the D3R crystal structure for rational drug design. We have taken the approach of deconstructing several D3R-selective substituted-4-phenylpiperazine antagonists and partial agonists into pharmacophoric elements. Using computational simulations and binding and activation studies, we have dissected the structural bases for D3R selectivity and efficacy. We find that selectivity arises from divergent interactions within a second binding pocket separate from the OBS, whereas efficacy depends on the binding mode in the OBS. Our findings reveal structural features of the receptor that are critical to selectivity and efficacy that can be used to design highly D3R-selective ligands with targeted efficacies. This drug design strategy has been undertaken to discover novel tools that have both high affinity and selectivity for the D3R, but also predicted efficacy, which until now has been elusive. In addition, we have made progress in discovering molecules that also have appropriate biopharmaceutical properties for in vivo studies that may provide leads toward efficacious medications to treat addiction.

Supported by NIDA-IRP (AHN), DA022413, MH54137 (JAJ), DA23957 and DA13584 (RRL), and DA023694 (LS).

Cue-Induced Dopamine Release in Striatal and Extra-striatal Regions in Cocaine Dependent Users: A High Resolution PET [18F]Fallypride Study

Aryandokht Fotros[1,2], **Kevin F. Casey**[1], **Kevin Larcher**[2], **Jeroen A.J. Verhaeghe**[2], **Sylvia M.L. Cox**[2], **Paul Gravel**[2,3], **Andrew J. Reader**[2,3], **Alain Dagher**[2], **Chawki Benkelfat**[1,2] and **Marco Leyton**[1,2]

[1]*Department Psychiatry, McGill University, Montreal, QC, Canada;* [2]*Department of Neurology & Neurosurgery, McGill University, Montreal, QC, Canada;* [3]*Department of Biomedical Engineering, McGill University, Montreal, QC, Canada*

BACKGROUND: Drug related cues are potent triggers for relapse in people with cocaine dependence. Dopamine release within a limbic network of striatum, amygdala and hippocampus has been implicated in animal studies, but in humans it has been possible to test the first region only.

OBJECTIVES: The objective was to measure cocaine cue-induced dopamine release in the amygdala and hippocampus using high-resolution PET with [18F]fallypride.

METHODS AND RESULTS: Twelve cocaine dependent volunteers (mean age: 39.6 ± 8.0; years of cocaine use: 15.9 ± 7.4) underwent two PET [18F]fallypride scans with a Siemens HRRT camera, one with exposure to neutral cues and one with cocaine cues. [18F]Fallypride non-displaceable binding potential (BP_{ND}) values were derived for five regions of interest (ROI) (ventral limbic striatum, associative striatum, sensorimotor striatum, amygdala, and hippocampus). Subjective responses to the cues were measured with visual analog scales and grouped using principal component analysis. Individual differences in the cue-induced craving factor predicted [18F]fallypride responses in the ventral limbic ($r = 0.581$, $p = 0.048$), associative ($r = 0.589$, $p = 0.044$) and sensorimotor striatum ($r = 0.675$, $p = 0.016$); the greater the craving, the greater the [18F]fallypride response. When participants were split into high ($n = 6$) vs. low craving responders ($n = 6$), drug cue exposure significantly decreased BP_{ND} values in the high craving group in all five ROI (limbic striatum: $p = 0.019$, associative striatum: $p = 0.008$, sensorimotor striatum: $p = 0.004$, amygdala: $p = 0.040$, and right hippocampus: $p = 0.025$), but not in the low craving group.

CONCLUSIONS: To our knowledge this study provides the first evidence of drug cue-induced dopamine release in the amygdala and hippocampus in humans. The preferential induction of dopamine release among high cue-responders suggests that these aspects of the limbic reward network might contribute to drug seeking behavior.

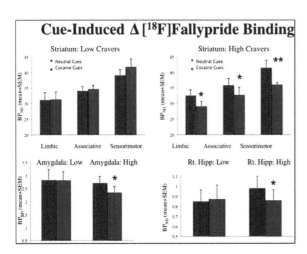

Cue-Induced Δ [18F]Fallypride Binding

β-arrestin Regulation of Ghrelin Signaling in modulating Addictive Behavior

Tama Evron, Nikhil M. Urs, Laurie Sutton, Yushi Bai, Marc G. Caron and Larry S. Barak

Department of Cell Biology, Duke University Medical Center, Durham, NC 27710, USA.

BACKGROUND: Over twenty millions individuals in the United States are frequent users of illicit drugs. The reinforcing behavior associated with drugs of abuse originates in dopaminergic neurons within the CNS. The endogenous GHSR1a ghrelin receptor appears to be one of the strongest modulators of CNS dopamine signaling underlying reward. GHSR1a is a GPCR, and like other GPCRs should signal through two distinct pathways, one regulated by G-proteins and the other by β-arrestins. Each signaling arm is thus a potential pharmacological target for either classical ligands or functionally selective biased ligands.

OBJECTIVES: We aim to develop selective ligand(s) for GHSR1a that will attenuate the reward properties of drugs of abuse with minimum side effects. For this aim, we are studying the role of β-arrestins vs. G proteins in the signaling of GHSR1a and in addictive behaviors by using mouse models of altered dopamine neurotransmission, β-arrestin KO mice and GHSR1a mutants.

METHODS AND RESULTS: The small molecule antagonist of GHSR1a, Yil781 reduced hyper-locomotion in the dopamine transporter (DAT) KO mice and in cocaine-treated WT mice in a dose dependent manner; validating that GHSR1a small molecule ligands can affect dopamine-mediated behaviors. In a cellular system, we show that in addition to its canonical activation of Gq, GHSR1a signals through β-arrestins. GHSR1a mutants can be biased toward either Gq or β-arrestin signaling, affecting downstream signaling through either ERK1/2 or the small GTPase RhoA.

Furthermore, stimulation of GHSR1a with the small molecule agonist L585 facilitates stress fiber formation in U2OS cells stably express GHSR1a, inhibited by pre-incubation with YIL781. This effect may suggest a potential mechanism for the reported drug-induced synaptic plasticity.

CONCLUSIONS: Altogether, we believe that the GHSR1a presents a promising pharmacological target for treating addiction and that β-arrestin signaling may play an important role in modulating dopaminergic signaling to this end.

GHSR1a, 366 aa

Gq Signaling

β-arrestin Signaling

GHSR1a-medidaited dopamine-dependent behaviors

Dopamine D3 and D4 Receptors as Therapeutic Targets for Nicotine Addiction

Bernard Le Foll[1], Maram Khaled[1], Yijin Yan[1], Yann Le Strat[1], Steven R Goldberg[2] and Abhiram Pushparaj[1]

[1]*Translational Addiction Research Laboratory, CAMH and University of Toronto, Ontario, Canada;* [2]*Preclinical Pharmacology Section, Behavioral Neuroscience Research Branch, NIDA-IRP, NIH, DHHS, Baltimore, MD, USA*

BACKGROUND: It has been suggested that the dopamine D3 receptor (DRD3) is involved in the mechanisms underlying stimulus-controlled drug-seeking behavior. In contrast, since cloning of the dopamine receptor D4 (DRD4), its role in the brain has remained unclear. It has been reported that polymorphism of the DRD4 gene in humans is associated with reactivity to cues related to tobacco smoking. However, the role of the DRD4 in animal models of nicotine addiction has seldom been explored.

OBJECTIVES: A DRD3 antagonist (SB 277011-A) and a selective DRD4 antagonist (L-745,870) were evaluated on models of nicotine addiction in rats.

METHODS AND RESULTS: SB 277011-A (1−10 mg/kg) blocked cue-induced reinstatement of nicotine-seeking. It has also been reported that this DRD3 antagonist reduces nicotine-induced reinstatement of nicotine-seeking, while nicotine-taking is not affected. Similarly, L-745,870 significantly attenuated reinstatement of nicotine-seeking induced by both nicotine-associated cues and by nicotine priming, but did not affect established nicotine self-administration behavior or reinstatement of food-seeking behavior induced by food cues or food priming.

CONCLUSIONS: As DRD3 and DRD4 blockade selectively attenuated both cue- and nicotine-induced reinstatement of nicotine-seeking behavior, DRD3 and DRD4 antagonists are potential therapeutic agents against relapse to tobacco smoking.

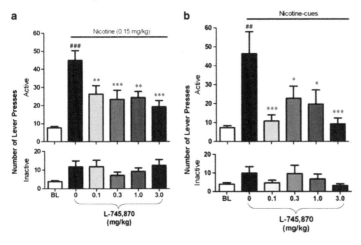

References

Yan, Y., et al., *Blockade of dopamine d4 receptors attenuates reinstatement of extinguished nicotine-seeking behavior in rats.* Neuropsychopharmacology, 2012. **37**(3): p. 685−96.

Khaled, M.A., et al., *The selective dopamine D3 receptor antagonist SB 277011−A, but not the partial agonist BP 897, blocks cue-induced reinstatement of nicotine-seeking.* Int J Neuropsychopharmacology, 2010. **13**(2): p. 181−90.

Long-Term Cocaine Exposure Affects Dopamine D_3 Receptor Function.

**Michael A. Nader[1], Angela Duke[1], Susan H. Nader[1], Paul W. Czoty[1],
Robert E. Brutcher[1] and Amy H. Newman[2]**

[1]*Department of Physiology and Pharmacology, Wake Forest University School of Medicine,
Winston-Salem, NC 27157;* [2]*Medicinal Chemistry Section, Intramural Research Program, National
Institute on Drug Abuse, Baltimore, MD 21224*

BACKGROUND: Dopamine receptors consist of two superfamilies, D1- and D2-like receptors. A subtype of the D2-like family, the D_3 receptor, has been a focus of addiction treatments because of the restricted distribution to limbic brain regions (Heidbreder and Newman, 2010).

OBJECTIVES: The behavioral effects of compounds that function, *in vitro*, as full agonists (quinpirole), partial agonists (CJB090, PG619), or antagonists (NGB2904, buspirone) at D_3 receptors were examined in several monkey models.

METHODS AND RESULTS: Unconditioned (drug-elicited yawning) and conditioned (cocaine self-administration, SA and drug discrimination, DD) behaviors were studied. Previously, we reported that quinpirole elicited yawns in cocaine-naive monkeys and cocaine-experienced monkeys (Martelle et al., 2007; Blaylock et al., 2011), but D_3 partial agonists only elicited yawns in monkeys with a cocaine history. D_3 partial agonists decreased quinpirole-elicited yawns, suggesting an antagonist effect, but did not reinstate cocaine seeking, while quinpirole did (Fig). In DD, PG619 does not substitute for quinpirole, but potentiates quinpirole's effects. PG619 did not selectively decrease cocaine SA relative to food reinforcement when responding was maintained under fixed-ratio or progressive-ratio reinforcement schedules and cocaine-food choice was not affected by 5 days of treatment. When cocaine SA occurred under a second-order schedule, PG619 decreased high rates of responding, but increased low rates of cocaine SA. The latter findings are reminiscent of clinical trials and suggest that individualized treatments would be the best strategy for cocaine abuse. To that end, we are combining PET imaging of D2-like and D_3 receptors with behavior to better determine which compounds may be effective in selectively decreasing cocaine SA.

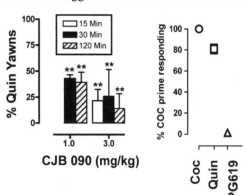

CONCLUSIONS: The continued investigation of D_3 partial agonists under multiple experimental conditions will lead to a better understanding of the neuropharmacological and behavioral consequences of long-term cocaine use, which will aid in the development of novel treatments for psychostimulant addiction. DA12460, DA25120, DA06634, NIDA-IRP.

Catecholamines in the Periphery

David Goldstein and Lee Eiden

The Theme on catecholamines in the periphery covered the several roles played by catecholamines in physiology and pathophysiology outside the brain. Norepinephrine (NE) is the main neurotransmitter of the sympathetic nervous system (SNS) mediating regulation of the circulation. Hypofunctional disorders of the sympathetic noradrenergic system include autonomic failure syndromes, and hyperfunctional disorders, postural tachycardia syndrome (POTS). Pheochromocytomas are rare but scientifically important tumors of catecholamine-synthesizing cells.

Freeman introduced novel methods of peripheral autonomic assessment, with emphasis on punch skin biopsies stained with specific neuronal markers. Immunohistochemical stains specific for sympathetic catecholaminergic innervation reveal that the majority of nerve fibers within arrector pili muscles are catecholaminergic. Pilomotor function can be assessed by the iontophoresis of pharmacologic agents that stimulate the muscles directly or via an axon reflex. Thus, skin biopsy and the associated functional assessments may be useful biomarkers providing diagnostic and prognostic information in autonomic disorders.

Raj et al. reviewed genetic and acquired determinants of POTS. POTS is best viewed as syndrome in which excessive orthostatic tachycardia can be a final common pathway for effects of a variety of pathophysiological processes, such as hypovolemia, mast cell activation, partial autonomic neuropathy, and central hyperadrenergic states. Identification of pathophysiological mechanisms in individuals will hopefully lead to more rational, pathophysiology-targeted treatments.

Robertson described the therapeutic value of the osmopressor response. In patients with impaired baroreflex function, water ingestion induces a robust increase in blood pressure and vascular resistance, a response termed the Osmopressor Response (OPR). The OPR does not occur with i.v. 5% glucose or i.v. physiological saline, and so the gastrointestinal tract or splanchnic circulation may be an important modulator of cardiovascular regulation. Blocking ganglionic transmission prevents the OPR, indicating that autonomic mechanisms mediate the OPR. Clinical studies provide evidence that increased SNS activity underlies the pressor effect of water. Biaggioni discussed autonomic contributions to sustained hypertension. Sympathetic activation was one of the first targets in the treatment of hypertension, but current antihypertensive drugs targeting sympathetic outflow are limited by side effects. This void is being filled by interventional approaches. Esler described a non-pharmacologic approach for treating clinical hypertension based on renal sympathetic denervation produced by renal artery radiofrequency ablation. Central sympathetic inhibition is observed as documented by peroneal sympathetic microneurography. Insulin sensitivity also increases post-procedure.

O'Connor reviewed studies about chromogranin A (CHGA) in blood pressure regulation via catecholamine systems. CHGA represents a control point in the formation of catecholamine storage vesicles, as well as the source of the peptide catestatin, an inhibitor of the physiological (nicotinic cholinergic) pathway for catecholamine release. Virtually all the variation at CHGA occurs as bi-allelic single nucleotide polymorphisms (SNPs). Amino acid replacement variation within the catestatin region results in Gly/Ser heterozygosity. Ser allele carriers had enhanced baroreflex sensitivity and diminished catecholamine secretion and seemed to be relatively protected from development of systemic hypertension.

Gharaibeh et al. reported on stress-induced hypertension, in caregiver spouses of Alzheimer's disease patients. Caregiving and obstructive sleep apnea were associated with increased blood pressure, beta-2 adrenoceptor desensitization, and elevated plasma NE levels, suggesting that stress increases blood pressure via increased SNS outflow associated with sleep disruption. Kvetnansky et al. found evidence that adipocytes possess catecholamine-synthesizing enzymes and that immobilization stress increases adipocyte catecholamine content. Santana described changes in mouse adrenal medulla upon exposure to chronic unpredictable stress.

Ehrhard-Bornstein pointed out that the adult adrenal medulla contains precursor cells thought to contribute to the gland's plasticity and adaptation to physiological needs, and the presence of these progenitor cells provides an extraordinary model to study the process of stem cell activation from quiescence to proliferation and differentiation into chromaffin and neural lineages. Basu proposed that DA, via D2 receptors, controls functions of tumor blood vessels, inhibiting tumor growth and potentially increasing efficacy of conventional anti-cancer treatment. They found evidence for increased adrenal medulla size and catecholamine content, associated with chromaffin progenitor cell proliferation.

La Gamma et al. reported evidence that recurrent hypoglycemia differentially regulates catecholamine release and gene expression. Recurrent hypoglycemia and the attendant defective counter-regulatory response are the main limiting factors for long-term health benefits of intensive insulin therapy in diabetes; a pathological condition known as hypoglycemia-associated autonomic failure (HAAF). La Gamma and colleagues found that depending on the frequency of the hypoglycemic episodes and the attendant increase in trans-synaptic adrenal cholinergic activity, recurrent hypoglycemia can either increase TH mRNA to sustain the releasable pool of EPI or, paradoxically, suppress TH mRNA, resulting in HAAF due to an inability to replenish releasable EPI stores.

Pheochromocytomas (PHEOs) are rare catecholamine-producing tumors most commonly located in the adrenal gland and less frequently in extra-adrenal locations (paragangliomas, PGLs). Lenders asked what is the optimal biochemical diagnostic strategy for pheochromocytoma. For patients in whom there is suspicion or risk of pheochromocytoma or paraganglioma, the recommendation is that initial biochemical testing include measurements of plasma free metanephrines or urinary fractionated metanephrines. Pacak pointed out that genetically linked PHEOs/PGLs are much more common than previously thought. Thirty-five percent of all these tumors are due to a genetic mutation, and 50% of extra-adrenal tumors (except head and neck PGLs) are caused by succinate dehydrogenase (SDH) gene mutations. For diagnostic evaluation by imaging, PET scanning should become a gold standard. ^{18}F-DOPA PET is best for head and neck PGLs and ^{18}F-dopamine for sympathetic (outside head and neck) PHEOs and PGLs, with the exception of SDHB related metastatic tumors where ^{18}F-fluorodeoxyglucose should be used first.

Recurrent Hypoglycemic Stress Differentially Regulates Catecholamine Release and Transmitter Gene Expression

Edmund La Gamma[1], Necla Kirtok[1], Owen Chan[2] and Bistra Nankova[1]

[1]*Division of Newborn Medicine, Departments of Pediatrics, Biochemistry and Molecular Biology, New York Medical College and The Regional Neonatal Center, Maria Fareri Children's Hospital at Westchester Medical Center-New York Medical College, USA;* [2]*Yale School of Medicine, Department of Internal Medicine, Division of Endocrinology and Metabolism, USA*

Recurrent hypoglycemia and the attendant defective counter-regulatory response are the main limiting factors for long-term health benefits of intensive insulin therapy in diabetes; a pathological condition known as hypoglycemia-associated autonomic failure (HAAF). The molecular mechanisms underlying the progressive loss of circulating epinephrine responses (a hallmark sign of HAAF) are largely unknown but contribute to clinical risks due to the additional problem of patient "unawareness." We found that the effects of three sequential, once daily hypoglycemic events resulted in an increase in plasma epinephrine with a corresponding up-regulation of adrenomedullary TH mRNA and enkephalin mRNA following a 90 min hypoglycemic clamp on day 4. In contrast, exposure to recurrent twice-daily hypoglycemia reduced the immediate counter- regulatory epinephrine responses on day 4 and suppressed the induction of TH mRNA yet further induced enkephalin mRNA. In vitro run-on assay evidence in PC12 cells supports accelerated degradation of TH mRNA. These results suggest that: 1) the transsynaptic transmission of the signal following twice daily episodes of recurrent hypoglycemia is intact; 2) Adrenal TH and enkephalin gene expression are subjected to differential regulation; and 3) Post-transcriptional mechanisms may contribute to the in vivo effects on TH mRNA. Glucagon plasma levels behaved similarly to epinephrine while corticosterone levels were elevated under all handling conditions as was PNMT mRNA. Subsequent studies using preferential, peripheral acting, nicotinic receptor (ganglionic) partial agonists attenuated the twice-daily effects. We conclude that depending on the frequency of the hypoglycemic episodes (and the attendant increase in transsynaptic adrenal cholinergic activity) recurrent hypoglycemia can either increase TH mRNA (once daily) to sustain biosynthesis and thus, the releasable pool of epinephrine or, paradoxically, suppress TH mRNA levels (twice daily) resulting in HAAF due to an inability to replenish releasable epinephrine stores. The evidence that attenuation of peripheral nicotinic receptor pathways preserves epinephrine biosynthesis is similar to functions of central-acting, clinically useful therapies (e.g. Chantix®); see poster by Nankova, et al. (this meeting). Exposing this dual regulatory mechanism may help explain the clinical phenomenon of HAAF and provides new opportunities to prevent it. Moreover, the persistent elevation of opiate neuropeptide gene expression under all conditions may have evolutionary significance as an endogenous analgesic in terminal illness.

Sympathetic Mechanism of Hypertension

Italo Biaggioni

Vanderbilt University School of Medicine, USA

The autonomic nervous system is not only crucial role in the instantaneous regulation of blood pressure, but also contributes to the chronic maintenance of hypertension, as evidenced by conditions resulting from lesions of autonomic pathways and recent findings in obesity- and resistant-hypertension. Lesions of baroreflex pathways in the neck (following surgery or radiation) or the NTS, lead to labile hypertension. Neurovascular compression of the RVLM is associated with hypertension. Neurodegeneration of central autonomic pathways (multiple system atrophy) is accompanied by severe supine hypertension driven by residual sympathetic tone. These rare disorders support the concept that abnormal autonomic mechanisms can contribute to the maintenance of hypertension. Obesity, the most common cause of hypertension, is characterized by selective activation of sympathetic pathways involved in cardiovascular regulation. Furthermore, blood pressure is virtually normalized in animal models of obesity hypertension with chronic carotid sinus stimulation and in patients by autonomic withdrawal with ganglionic blockade. It was thought that sympathetic activity is increase in obesity as a compensatory mechanism to increase energy expenditure. However, clinical studies have shown that blood pressure is virtually normalized by ganglionic blockade in obese hypertensives, but the elevated resting energy expenditure associated with obesity is not. Furthermore, sympathetic activation may contribute to insulin resistance. Sympathetic activation was one of the first targets in the treatment of hypertension, but current antihypertensives aiming the autonomic nervous system are limited by side effects. This void is being filled by interventional approaches such electrical stimulation of the carotid sinus, and catheter ablation of renal afferent nerves. These novel devices are currently being tested for the treatment of resistant hypertension.

References

1. Shibao C, Gamboa A, Diedrich A, Ertl AC, Chen KY, Byrne DW, et al. Autonomic contribution to blood pressure and metabolism in obesity. HTN. 2007;49(1):27−33.
2. Esler MD, Eikelis N, Lambert E, Straznicky N. Neural mechanisms and management of obesity-related hypertension. CurrCardiolRep. 2008;10(6):456−63.
3. Esler MD, Krum H, Sobotka PA, Schlaich MP, Schmieder RE, Bohm M. Renal sympathetic denervation in patients with treatment-resistant hypertension (The Symplicity HTN-2 Trial): a randomised controlled trial. Lancet. 2010 Dec 4;376(9756):1903−9.
4. Lohmeier TE, Iliescu R, Liu B, Henegar JR, Maric-Bilkan C, Irwin ED. Systemic and renal-specific sympathoinhibition in obesity hypertension. Hypertension. 2012 Feb;59(2):331−8.
5. Biaggioni I. Interventional approaches to reduce sympathetic activity in resistant hypertension: to ablate or stimulate? HTN. 2012 Feb;59(2):194−5.

Harnessing the Autonomic System for Therapeutic Intervention

Murray Esler

Baker IDI Heart and Diabetes Institute, Australia

Pharmacological targeting of the autonomic nervous system has a long history. A recent initiative, differing from this, has been the development of devices for the management of drug-resistant hypertension. These modify the sympathetic nervous system reversibly with one device (the implantable arterial baroreceptor stimulator) and permanently with the other (the radiofrequency renal nerve ablation catheter). Studies are currently being conducted to evaluate the efficacy and safety of the barostimulation procedure, and to identify the hypertensive population who might receive greatest benefit. With endovascular renal denervation, documentation of increased renal sympathetic outflow in essential hypertension (with the isotope dilution, norepinephrine spillover technique), coupled with BP lowering from renal surgical denervation in experimental hypertension models, provided the theoretical basis for treatment of drug-resistant essential hypertension in this way. Laparoscopic surgery is not feasible; as the nerves lie in the renal artery wall, bilateral renal artery transection and re-anastomosis would be needed. My colleagues and I, working with ARDIAN corporation USA, developed and tested an alternative, and successful technique, utilising intra-luminal radiofrequency energy delivery, which has been shown to safely denervate the kidneys, and to lower blood pressure in patients with previously uncontrollable essential hypertension (Lancet 2010; 376:1903-1909). It was anticipated that efferent renal sympathetic denervation would fully explain the antihypertensive effect of the procedure. Central sympathetic inhibition, however, is observed (documented by sympathetic nerve recording using microneurography), an effect arising from ablation of renal afferent nerves projecting to the CNS (N Engl J Med 2009;361:932-934). Also surprising was an increase in insulin sensitivity post-procedure, confirmed with the glucose clamp procedure, possibly attributable to the central sympathetic inhibition from renal deafferentation (Circulation 2011;123:1940-1946). The durability of the BP lowering (with follow-up now beyond 3 years) and the safety of the procedure now justifies evaluation of whether a "cure" might be achieved, by applying the technique in patients with milder grades of hypertension. Studies of this type have commenced. A recent development has been the setting up of clinical trials to evaluate whether endovascular renal denervation might be of therapeutic benefit in patients with renal hypertension, or in congestive heart failure. In renal hypertension activation of the sympathetic nervous system is at a high level, and catheter-based renal denervation has been demonstrated in a pilot study to potently reduce blood pressure (D Hering et al, J Am Soc Nephrol, published online 17 May 2012). Earlier investigation of the neural pathophysiology of heart failure, which demonstrated preferential activation of the cardiac sympathetic outflow (Circulation 1986;73:615-621) provided the theoretical basis for the successful clinical evaluation of beta-adrenergic blockade

(N Engl JMed 1996;334:1349-1355). Now in the beta-blockade era, with the risks of cardiac sympathetic activation being adequately neutralized, the level of renal sympathetic activation in heart failure patients, assessed with isotope dilution measurements of renal norepinephrine spill-over, has emerged as the new clinical predictor of clinical outcome. Therapeutic targeting of this renal sympathetic activation with renal denervation is a logical next step, and clinical trials have commenced.

Norepinephrine Transporter Function and Human Cardiovascular Regulation

Jens Jordan

Hannover Medical School, Germany

Approximately 80-90% of the norepinephrine released in the brain or in peripheral tissues is taken up again through the neuronal norepinephrine transporter (NET). Norepinephrine is a crucial cardiovascular neurotransmitter and perturbed norepinephrine turnover can cause or exacerbate cardiovascular disease. Yet, NET received surprisingly little attention in human cardiovascular research. Genetic NET dysfunction is a rare cause of the postural tachycardia syndrome. The condition is characterized by excessive adrenergic stimulation of the heart, particularly with standing. Pharmacological NET inhibition in healthy subjects reproduces the clinical phenotype. Studies with NET inhibitors also showed that NET has opposing effects on cardiovascular sympathetic regulation in the brain and in the periphery. For example, NET inhibition acutely attenuates the cold pressor response. Furthermore, NET is involved in the distribution of sympathetic activity between vasculature, heart, and kidney. Redistribution of sympathetic activity towards the heart may explain why NET inhibitors prevent neurally mediated syncope during head-up tilt, which results from acute sympathetic withdrawal. Biochemical studies suggested reduced NET function in some patients with essential hypertension. Furthermore, cardiac NET function appears to be reduced in common heart diseases, such as congestive heart failure, ischemic heart disease, and stress-induced cardiomyopathy. Whether NET dysfunction is a consequence or cause of progressive heart disease is not understood. Given the widespread use of medications inhibiting NET, these issues ought to be studied in more detail.

Role of Common Genetic Variation in Control of Human Adrenergic Function and Blood Pressure: Focus on the Chromogranins

Daniel O'Connor
UCSD, USA

RATIONALE: CHGA represents a control point in the formation of catecholamine storage vesicles, as well as the source of the peptide catestatin (human CHGA[352-372], an inhibitor of the physiological (nicotinic cholinergic) pathway for catecholamine release. Given these physiological roles, we asked whether naturally occurring genetic variation at the human CHGA locus influenced human autonomic and disease traits.

METHODS: The CHGA locus was thus re-sequenced for systematic polymorphism discovery in n = 180 individuals (i.e., n = 360 chromosomes), allowing discovery of variants down to $\sim 0.3\%$ minor allele frequency (MAF). We then tested whether such variants predicted human autonomic or cardio-renal disease traits. Finally, we tested the function of allelic variants. chromaffin cell-transfected luciferase reporter plasmids

RESULTS: Virtually all of the variation at CHGA occurred as bi-allelic SNPs (single nucleotide polymorphisms) Within the open reading frame (ORF), non-synonymous (amino acid replacement) variation within the catestatin region (at Gly364Ser, rs9658667, MAF $\sim 3\%$), resulted in Gly/Ser heterozygosity in $\sim 6\%$ of the European ancestry population. Ser allele carriers displayed enhanced baroreceptor sensitivity, diminished pressor responses to environmental (cold) stimulation, and diminished catecholamine secretion. Ser allele carriers (especially men) also seemed to be relatively protected from development of systemic hypertension, in case/control studies. Gly364Ser differentially influenced catecholamine release from chromaffin cells and neurons in culture. In the 3'-UTR, common variant C + 87T (rs7610, MAF $\sim 27\%$) was associated with diminished pressor responses, as well as risk of developing hypertension, especially in males. The same variant was associated with risk of developing hypertensive renal failure in African-Americans, as well as rate of loss of renal function in progressive renal disease. Finally, C + 87T proved to be differentially functional in chromaffin cells transfected with luciferase/3'-UTR reporter plasmids. In the proximal promoter, 8 common SNPs were discovered in the first ~ 1.2 kbp upstream of Exon-1. Promoter haplotypes predicted the BP response to environmental stress. Promoter haplotype/luciferase reporter transfections established a prominent functional role for variant G-462A (rs9658634, MAF $\sim 21\%$), which disrupted at least one transcriptional control motif.

CONCLUSIONS: Naturally occurring, common genetic variation at CHGA has functional consequences, and predisposes to cardio-renal disease risk. ACKNOWLEDGEMENTS: Critical participants in these studies, without whom the work would not be possible, include Gen Wen, Bruce A. Hamilton, Fangwen Rao, Yuqing Chen, Kuixing Zhang, Rany M. Salem, Sushil K. Mahata, Manjula Mahata, Stephane Chiron, Jiaur Gayen, Sucheta Vaingankar, Sajalendu Ghosh, Mats Stridsberg, Barry Freedman, Maple M. Fung, Michael G. Ziegler, and Nicholas J. Schork.

What is the Optimal Biochemical Diagnostic Strategy for Pheochromocytoma?

Jacques Lenders

Department of Internal Medicine, Radboud University Nijmegen Medical Centre, Nijmegen, The Netherlands;
Department of Internal Medicine III, University Hospital Gustav Carus, Dresden, Germany

Pheochromocytoma and extra-adrenal paragangliomas are rare catecholamine−producing tumors which, if not timely and properly diagnosed, may result in potentially fatal cardiovascular accidents. Presenting clinical signs and symptoms are quite aspecific, extremely variable and usually paroxysmal in nature. These are the main reasons why its diagnosis suffers from long delays. The pivotal step in reducing the rate of delayed or missed diagnoses is early consideration of this tumor in the differential diagnosis. Knowledge of the available modern biochemical tests to demonstrate catecholamine excess production is indispensable to choose the most appropriate biochemical test. For daily clinical practice the optimal test for initial screening is a test with a nearly 100% negative predictive value in conjunction with an as highly as possible positive predictive value. The two most solid assay methods for measurement of fractionated metanephrines are HPLC with electrochemical detection and liquid chromatography with tandem mass spectrometry (LC-MS/MS). Employing proper sampling conditions and use of appropriate reference intervals is critical for correct interpretation of test results. Currently there is firm evidence that measurement of fractionated metanephrines, either in urine or plasma, offers tests with 96-99% sensitivity and a concomitant specificity of 45-96%, depending on whether it is a sporadic or a hereditary pheochromocytoma. So a normal test result reliably excludes the presence of a catecholamine-producing tumor. Conversely, an increased test result does not always proof the presence of a tumor. An important source of false-positive test results is the use of all kinds of medication, in particular antidepressants. To eliminate persistent false-positive test results, even after stopping interfering medication, a clonidine suppression test (using plasma normetanephrine measurement) can be used in patients who do show slightly elevated levels of plasma free metanephrines (<4x the upper reference level). In conclusion, for patients in whom there is suspicion or risk of pheochromocytoma or paraganglioma, we recommend that initial biochemical testing should include measurements of plasma free metanephrines or urinary fractionated metanephrines since, when correctly implemented, these measurements represent the most accurate of all currently available biochemical tests for diagnosis of the tumors.

SDHx-Related Pheochromocytoms: New Advances in Genetics and Functional Imaging

Karel Pacak
NIH, USA

BACKGROUND: Pheochromocytoma (PHEO) are catecholamine producing tumors that are most commonly located in the adrenal gland and less frequently in extra-adrenal locations (so called paragangliomas /PGL/).

OBJECTIVES: Recently, several new discoveries have been made that have substantially improved approaches and strategies in the pathogenesis, genetics, diagnosis, localization, and treatment of these tumors.

METHODS AND RESULTS: Genetically linked PHEOs/PGLs are much more common than previously thought. Thirty-five percent of all tumors are due to a genetic mutation, and 50% of extra-adrenal tumors (except head and neck PGLs) are caused by succinate dehydrogenase (SDH) gene mutations. Recently, some gastrointerstinal stromal tumors so called GIST have been found to be associated with the presence of succinate dehydrogenase subunit B/D (SDHB/D) gene mutation and paraganglioams. The loss of immunohistochemistry for SDHB or SDHA, respectively was found to "mark" the presence of SDHB/C/D or SDHA related PHEOs and to use it to triage a proper genetic testing. New gene, carboxypeptidase A, has been implicated in the pathogenesis and prediction of metastatic or recurrent disease in some of these tumors. SDHB gene mutations have been implicated as the most common cause of malignant PHEOs and PGLs in both children and adults. In children and adolescents, new study showed that familial tumors (especially SDHB and SDHD) may be twice as common as in adults. Mediastinal, Zuckerkandl organ and cardiac PGLs were found to be most commonly related to SDHx gene mutations. It has been proposed that all patients with metastatic PHEOs and PGLs should undergo first SDHB genetic screening; those with multiple PGLs including head and neck tumors should have first SDHD gene screening. Children as SDHB gene mutation carriers should be screened starting at age 5 with plasma metanephrines to be performed initially. Functional imaging using positron emission tomography compounds should become a gold standard in the localization of these tumors. [18F]-fluorodopa PET is the best imaging modality for head and neck PGLs, [18F]-fluorodopamine PET is the best for sympathetic (outside head and neck) PHEOs and PGLs, with the exception of SDHB related metastatic tumors where [18F]-fluorodeoxyglucose

PET should be used as the first imaging modality. [123I]-MIBG is not recommended for the initial localization of these tumors, except adrenal PHEOs (where all but Octreoscan PHEO imaging methods perform similar) and situations when [131I]-MIBG treatment is considered (e.g. metastatic disease). PET/MRI is becoming a new imaging modality in the evaluation of these tumors, especially for bone and liver metastatic lesions after a radiofrequency ablation procedure.

CONCLUSIONS: Uncovering SDHx tumor-related pathways and their interactions together with the use of new molecular imaging to monitor cellular processes will facilitate new treatment options for these tumors and initiate truly personalized medicine in the near future.

Acknowledgements This research was supported by the Intramural Research Program of the Eunice Kennedy Shriver NICHD/NIH.

Novel Methods Peripheral Autonomic Assessment

Roy Freeman

MD Director, Center for Autonomic and Peripheral Nerve Disorders Beth Israel Deaconess Medical Center, Boston, MA, USA; Professor of Neurology Harvard Medical School, Boston, MA, USA

The skin, which contains sensory and autonomic fibers, is an accessible site for assessment of peripheral autonomic sudomotor, vasomotor and visceromotor structure and function, and provides the basis for several novel autonomic measures. The punch skin biopsy, stained with specific neuronal markers, can be used to assess sensory innervation and the innervation of organelles in the dermis including hair follicles, blood vessels, sweat glands and arrector pili muscles. These structural assessments can be complemented by parallel functional assessments. The skin also has the potential for yielding biomarkers that may provide diagnostic and prognostic information. The sympathetic cholinergic, sympathetic adrenergic and sensory nerve fibers that surround sweat gland tubules can be quantified with immunostaining. Nerve fibers that innervate sweat gland tubules can be stained with the pan-axonal marker protein gene product (PGP) 9.5 and imaged by light and confocal microscopy. These nerve fibers are primarily sympathetic cholinergic with some sympathetic adrenergic and sensory fibers. Several sudomotor functional tests exist to supplement the structural measures. These include assessment central and peripheral sudomotor system. The central sudomotor system is assessed with the thermoregulatory sweat test and peripheral system with the quantitative sudomotor axonal reflex test (QSART), the silastic imprint method and the quantitative direct and indirect reflex test (QDIRT)]. Blood vessels, which are present throughout the dermal layer in vascular networks within and around other dermal organelles, have sensory, adrenergic and cholinergic innervation that can be demonstrated with immunostaining. Vasomotor function can be assessed using the laser Doppler technique combined with physical, pharmacological and psychological stressors. Nerve fibers travel in parallel with the pili muscle that anchor the shaft of the hair follicle to dermal tissue. These fibers are easily identified through either light or confocal microscopy using the pan-axonal marker, such as PGP 9.5. Selective immuno-histochemical stains specific for sympathetic adrenergic innervation reveal that the majority of nerve fibers within pili muscles are sympathetic adrenergic. Pilomotor function can be assessed by the iontophoresis of pharmacologic agents that stimulate the muscles directly or via an axon reflex. Skin biopsy and the associated functional assessments may be useful biomarkers providing diagnostic and prognostic information in autonomic disorders such as diabetic neuropathy, amyloid neuropathy, toxic neuropathies, familial dysautonomia, Ross syndrome, Parkinson's disease, multiple system atrophy and cold induced sweating. These structural and functional assessments and their utility will be discussed.

Identification and Functional Characterization of Genetic Variants of the Catecholamine Release-Inhibitory Peptide Catestatin in an Indian Population

Bhavani S. Sahu[1], Lakshmi Subramanian[1], Prasanna K.R. Allu[1], Giriraj Sahu[1], Jagan Mohan[1], Pradeep K. Singh[2], Balashankar Gomathi[3], Samir K. Maji[2], Amal K. Bera[1], Sanjib Senapati[1], Ajit S. Mullasari[3] and Nitish R. Mahapatra[1]

[1]Department of Biotechnology, Indian Institute of Technology Madras, Chennai, India; [2]School of Bioscience and Bioengineering, Indian Institute of Technology Bombay, Mumbai, India; [3]Institute of Cardiovascular Diseases, Madras Medical Mission, Chennai, India

Catestatin (CST), a chromogranin A (CHGA)-derived peptide, is a potent endogenous inhibitor of the neuronal nicotinic acetylcholine receptor (nAChR)-activated catecholamine secretion. Mice lacking catestatin (generated by systemic deletion of the parent protein CHGA) display severe hypertension that can be rescued by administration of human CST. Consistently, the circulating CST level is diminished in hypertensive patients. Two naturally-occurring variants of this peptide (viz. Gly364Ser and Pro370Leu) have been detected in people of European ancestry. The status of genetic variants of CST in other human populations remains unknown. We aimed to identify and carry out functional analysis of naturally-occurring variants of the CST peptide in an Indian population. We re-sequenced the catestatin region of *CHGA* in an Indian population (n = 1000) and detected two amino acid substitution variants: Gly364Ser and Gly367Val; the 364Ser variant occurred at ∼2.8-fold higher frequency (as compared to the European population) while the 367Val variant is a novel one. Synthesized CST variant peptides (viz. CST-364Ser and CST-367Val) displayed 2.5- and 3.2-fold less potency (as compared to the wild-type peptide

[CST-WT]), respectively, for inhibition of nicotine-stimulated catecholamine secretion from PC12 cells. Consistently, patch-clamp experiments on PC12 cells revealed that the rank-order of blockade of acetylcholine-evoked inward current was: CST-WT > CST-364Ser > CST-367Val. In corroboration, the extents of blockade of nAChR-mediated intracellular $[Ca^{2+}]$ rise by these CST peptides were in the same order. To better understand the mechanistic basis for the differential activities of these CST peptides, we analyzed the structures of these peptides by CD spectroscopy and molecular dynamics simulation studies. The order of alpha-helical content in these peptides was: CST-WT > CST-364Ser > CST-367Val. Docking of the CST peptides onto a homology-modeled human $\alpha_3\beta_4$ nAChR structure (generated by using the 4 Å resolution structure of *Torpedo* nAChR), followed by molecular dynamics simulations showed that the rank order of binding affinity of these peptides with this nAChR subtype as well as the extent of occlusion of the receptor pore was: CST-WT > CST-364Ser > CST-367Val. In conclusion, our multi-disciplinary (biochemical, biophysical, cell biology, electrophysiology and computational biology) study unraveled that structural differences among the CST variants in an Indian population alter their binding affinity with nAChR, and thereby cause alterations in the extent of entry of Na^+ and Ca^{2+} into the cytosol, which in turn lead to differential inhibition of catecholamine release. These findings provide novel insights into understanding the nicotinic cholinergic signaling in humans and have implications for disease states wherein catecholamine levels in the circulation are dysregulated.

The Osmopressor Response in Human Physiology

David Robertson
Vanderbilt University, USA

Human subjects with impaired baroreflex function cannot buffer rises or falls in blood pressure (BP), thus allowing BP effects of endogenous or environmental stimuli that previously escaped detection to emerge dramatically. Even mild perturbations of autonomic control can impair the critical balance of blood pressure regulation in susceptible patients. Studies in such patients led us to discover that water ingestion induced a robust increase in BP and vascular resistance, a reponse we have termed the Osmopressor Response (OPR). The OPR is an increase in blood pressure in response to oral ingestion of pure water. The effect is especially large in patients with impaired baroreflex function, in whom oral ingestion of 16 oz (473 ml) water can raise systolic blood pressure (SBP) an average of 35 mmHg, with occasional patients having pressor surges of as much as 100 mmHg. This pressor response does not occur with intravenous 5% glucose or intravenous physiological saline. Thusit seems likely that the gastrointestinal tract or splanchnic circulation may be an important modulator of cardiovascular regulation. Administration of the nicotinic NNantagonist trimethaphan to block ganglionic transmission blocked the OPR, indicating that autonomic mechanisms are operative in eliciting the OPR. Clinical studies provide evidence that increased

Adrenergic Derived Myocardium: Anatomic Substrate for Stress Induced Cardiomyopathies

Candice Baker, Chaunhi Van Gregory Goldblatt, Kingsley Osuala and Steven N. Ebert

University of Central Florida College of Medicine, Orlando, FL 32827

BACKGROUND: Tako-Tsubo Syndrome is a stress-induced cardiomyopathy also known as "Broken Heart" Syndrome because it typically occurs following sudden emotional trauma that precipitates symptoms similar to those observed in myocardial infarction patients, but without significant coronary occlusion. Interestingly, Tako-Tsubo patients are often postmenopausal and present with severe chest pain due to hypo- or akinetic left ventricular (LV) function especially in the apical, mid, and sometimes basal regions of the LV. In addition, high levels of plasma catecholamines have been reported for Tako-Tsubo patients. We recently showed that adrenergic-derived cardiac cells become predominantly localized to the left side of the adult mouse heart (Osuala et al., PLoS ONE 6:e22811, 2011). Although it is well established that the adult heart expresses the adrenergic biosynthetic enzyme, phenylethanolamine n-methyltransferase (Pnmt), in the left atrium (LA) and LV, our study was the first to identify the specific anatomical distribution of adrenergic-derived cells in the adult heart. One of the striking features of this distribution pattern in the LV is the finger-like projections of concentrated adrenergic-derived myocardium in the basal, mid, and apical regions of the LV. The staining observed in the apical and mid-myocardial regions, in particular, bear strong resemblance to the regions of the heart primarily affected in Tako-Tsubo Syndrome patients.

OBJECTIVE: Our overall objective is to test the hypothesis that adrenergic-derived myocardium serves as an anatomical substrate for stress-induced cardiomyopathies such as Tako-Tsubo Syndrome.

METHODS AND RESULTS: To evaluate adrenergic-derived cells in the adult heart, we crossed Pnmt-cre knock-in mice with the LacZ ROSA 26 Reporter (R26R) strain, and maintained long-term colonies of the resulting offspring. Visualization of Pnmt + cells and their descendants was performed using a combination of XGAL and cell type-specific immunofluorescent histochemical staining. Thus far, we have performed histological assessments at 3, 6, and 9 months of age. Another group will be assessed at 12 months. In parallel, we have also collected LV physiological data in these mice using echocardiography techniques. We are experimenting with different pharmacological stress challenges and have performed a preliminary comparison of LV changes in normal adult male and female mice before and after administration of a single high dose epinephrine (50 mg/kg, i.p.) while under isoflurane anesthesia during real-time echocardiographic assessment with a high-resolution Vevo2100 ultrasound instrument (Visualsonics, Inc.). As expected, there was an overall increase in heart rate and cardiac output in both males and females in response to epinephrine, but the females were significantly more refractory to epinephrine stimulation than the males ($p < 0.05$, n = 30 per group). A subset of females from this group is currently undergoing induced menopause, and the stress tests will be repeated in these subjects when completed. The results will be compared to control pre-menopausal and male mice to determine if there is a difference in LV function between these groups. Upon sacrifice, hearts and adrenal glands are collected for histological staining as described above.

INTERPRETATION AND CONCLUSIONS: Preliminary conclusions are that both male and female mice display extensive swaths or bands of adrenergic-derived myocardium, and there are significant differences in LV responsiveness to epinephrine stress challenge in adult male vs. female mice at 9 months. Ongoing efforts are focused on re-evaluation of the epinephrine stress test following induction of menopause.

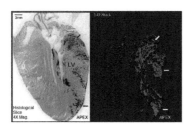

The Relationship of Norepinephrine to Stress Induced Hypertension

Munir Gharaibeh*, Ping Sun, Milos Milic and Michael G. Ziegler

*U. of Jordan, Amman, Jordan and U. of California, San Diego

We studied three groups to investigate how stress might induce hypertension. The first group was 144 subjects caring for a demented spouse with Alzheimer 's disease (AD) and 47 similar non-caregiving controls. Over 6 years, the hazards for hypertension were greater for caregivers than for controls (Cox Proportional Hazards, chi^2 [1, N = 174] = 4.86, p = 0.03). In a substudy, 37 elderly spousal caregivers and matched non-caregiver controls (mean age 73 years) were evaluated for life stress based on the Psychiatric Epidemiological Research Inventory and the Life Events and Difficulties Schedule. Those with high life stress had higher plasma norepinephrine (NE) levels (p < .04), which was related to down-regulated beta receptors on their circulating blood mononuclear cells. We evaluated sleep by full night home polysomnography in 40 spousal caregivers because Alzheimer's patients are often awake and disruptive at night. Among these caregivers, increased wake after sleep onset was positively associated with NE levels (beta = .35; t = 2.45, df = 32, p = .020). Overall, caregivers had disturbed sleep, even worse sleep after spousal death, and an increased incidence of new-onset hypertension. Caregivers who reported the greatest life stress had increased NE levels that were high enough to down-regulate their beta receptors.

The second group had obstructive sleep apnea (OSA) which is strongly associated with hypertension. Subjects received 2 weeks of continuous positive airway pressure (CPAP) versus sham-CPAP versus supplemental nocturnal oxygen. Plasma NE levels measured every 2 hours for 24 hours remained elevated at all time points in OSA. Treatment with CPAP returned NE to normal and lowered daytime BP by 2/2 mm Hg (p < .01) and nighttime BP by 6/4 mm Hg (p < .01) and returned beta receptor sensitivity from depressed levels.

In a third group, 784 community-dwelling, ambulatory men $\geq$ 65 years of age (mean age: 75.1 ± 4.9 years) did not have hypertension at the time of their in-home sleep studies (2003-2005) and returned for follow-up (2007-2009). Over a mean follow up of 3.4 years, lack of slow wave sleep (deep sleep) was significantly associated with the conversion of 243 of 784 men from normotension to hypertension.

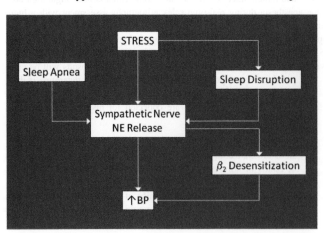

Stress from caregiving and from OSA was associated with increased BP, β_2 adrenergic receptor desensitization and increased NE. Our findings are consistent with the construct that chronic stress deprives persons of deep sleep. This is associated with increased NE levels and increased BP.

Regulation of Adrenal Phenylethanolamine N-methyltransferase Gene Expression and Adrenaline Synthesis in a Fetal Programming Model of Hypertension

Sandhya Khurana[1], Julie Grandbois[2], Phong Nguyen[2], Heather Peltsch[2], Suet Lo[3], Venkataraman Krishnan[1] and T.C. Tai[1,2,3,4]

[1]*Medical Sciences Division, Northern Ontario School of Medicine;* [2]*Dept.of Biology;* [3]*Dept. of Chem. and Biochem;* [4]*Biomolecular Sciences Program, Laurentian University, Sudbury, ON, Canada*

BACKGROUND: Human health and disease are influenced not only by the genetic make up of an individual, but also to a great degree by environmental factors. The prenatal environment can be a significant determinant of long-term health outcomes. An adverse fetal milieu such as undernourishment or exposure to environmental insults can have long-term consequences impacting adult health, a phenomenon known as fetal programming. Epidemiologically, *in utero* conditions have been linked to the development of diseases such as hypertension, diabetes and other pathophysiological conditions in adulthood. Studies suggest that fetal programming of adult diseases is

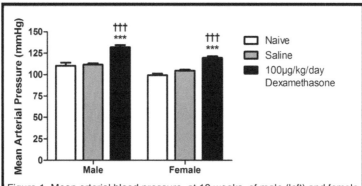

Figure 1. Mean arterial blood pressure, at 18 weeks, of male (left) and female (right) pups born to WKY dams injected with Dexamethasone (DEX) or saline during pregnancy. Error bars show SEM; significance between DEX and naive is denoted by * and between DEX and saline is denoted by†, p < 0.0001 (ANOVA followed by Newman - Keuls posthoc test).

mediated by glucocorticoids; either endogenous (eg. maternal stress), or exogenous (eg. synthetic glucocorticoid administered to aid in the development of the premature babies). Glucocorticoids regulate catecholamine biosynthesis and are critical for blood pressure homeostasis, with elevated levels leading to hypertension. The purpose of this study was to examine whether prenatal exposure to elevated glucocorticoid levels influences the development of adult hypertension via altered regulation of phenylethanolamine N-methyltransferase (PNMT), a candidate gene for hypertension and the terminal enzyme in the catecholamine biosynthetic pathway responsible for adrenaline biosynthesis.

OBJECTIVES: To understand how stress experienced during fetal life can be an antecedent for the development of adult hypertension and thereby increased cardiovascular risk, with an emphasis on the mechanisms by which *in utero* exposure to elevated levels of stress hormones (glucocorticoids) regulates PNMT expression.

METHODS AND RESULTS: Pregnant Wistar-Kyoto dams, were injected with 100 µg/kg/day of the synthetic glucocorticoid dexamethasone (DEX) in the third trimester. Blood pressure and weights of the offspring were measured from week 3-18, at which point the animals were sacrificed and tissues collected. Total RNA and protein were extracted from the adrenals and the expression of PNMT and its regulatory transcription factors analyzed by qPCR and Western blot. Plasma adrenaline was quantified by ELISA. Results from this study reveal that *in utero* exposure to DEX promotes the development of hypertension in the offspring, with males showing a more pronounced increase in blood pressure than females (Figure 1). Plasma adrenaline levels were also elevated in adult rats exposed to DEX. The mRNA transcripts for PNMT, and its transcriptional regulators Egr-1, Sp1, GR and AP-2, were elevated in the adrenal gland of these animals compared to saline controls, with differences in both males and females.

CONCLUSIONS: These results suggest that prenatal glucocorticoid exposure increases adrenal PNMT gene expression via altered transcriptional regulatory mechanisms. The influence of prenatal gluococorticoid stress on the programming of hypertension addresses key questions in understanding the origins of hypertension. The study emphasizes that the tightly regulated catecholamine biosynthesis can be disarrayed by *in utero* insults, resulting in elevated circulating catecholamines and subsequent hypertension in adulthood.

Chronic Unpredictable Stress Induces Catecholaminergic System Changes in Mouse Adrenal Gland

Magda M. Santana[1,2], **Vera Cortez**[1], **Célia Aveleira**[1], **Manuella Kaster**[1], **Joana Rosmaninho-Salgado**[1] and **Cláudia Cavadas**[1,2]

[1]*CNC-Center for Neuroscience and Cell Biology, University of Coimbra, Portugal;* [2]*Faculty of Pharmacy, University of Coimbra, Portugal.*

BACKGROUND: Prolonged exposure to stress leads to adaptations (or maladaptations) in adrenal medullary catecholamine content and catecholamine biosynthetic enzymes, which may underlie the onset of stress-related disorders. Nevertheless, the effect of chronic unpredictable stress (CUS) in adrenal medulla it still not completely understood.

OBJECTIVES: To study the effect of CUS on adrenal medullary system.

METHODS AND RESULTS: Adrenals obtained from mice submitted to 21 days of CUS (adapted from Lu *et al* 2006) were collected and weighted. The levels of catecholamine (CA) were quantified by HPLC-ED. The volumes of adrenal medulla and cortex were determined by the Cavalieri method and we observed that CUS induced an increase in adrenal cortex and adrenal medulla volumes. The total protein levels and the adrenal CA content of adrenal medulla were also increased in these animals. However, the amount of CA/mg of protein is lower in adrenal medulla from those animals. Moreover, alterations of tyrosine hydroxylase (TH) and phenylethanolamine-N-methyl transferase (PNMT) were also observed in adrenal medulla from stressed mice. To investigate whether CUS change cell proliferation and/or putative alterations in the adrenal gland progenitor cells, the levels of proliferating cell nuclear antigen (PCNA) and SOX9 were determined. The adrenal medulla from stressed animals have higher levels of protein levels of SOX9 and PCNA, compared to control, suggesting that chromaffin progenitor cell proliferation might be related with the increase in adrenal medulla size.

CONCLUSIONS: CUS induced an increase in adrenal medulla size and CA content, suggesting an adaptation of adrenal medulla function after prolonged exposure to unpredictable stressors. This adaptation might be related to changes in progenitor cell proliferation.

Supported by FCT and FEDER: PTDC/SAU-NEU/108110/2008, SFRH/BPD/31547/2006, SFRH/BD/44664/2008

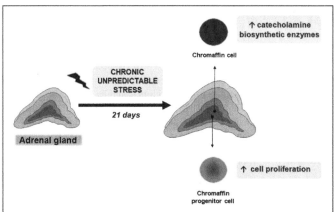

Reference

Lu et al. Proc Natl Acad Sci U S A. 2006 Jan 31;103(5):1593–8

Effect of Daidzein on Mean Arterial Pressure and Vascular Dysfunction in CLP-Induced Septic Mice

Thakur Uttam Singh, Somnath Bhojane, Subhashree Parida, Soumen Choudhury and Santosh Kumar Mishra

Division of Pharmacology and Toxicology, Indian Veterinary Research Institute, Izatnagar, Bareilly, Uttar Pradesh, India

BACKGROUND: The purpose of the study was to examine the effects of in vivo administration of daidzein on mean arterial pressure and vascular dysfunction in sepsis induced by caecal ligation and puncture (CLP). Sepsis is not only infection itself but also host response towards the tissue injury caused by infections.

OBJECTIVES: The aim of study was to investigate whether daidzein can restore mean arterial pressure and recover the vascular dysfunction.

METHODS AND RESULTS: Mice were treated with daidzein 10 mg/kg body weight intraperitonealy two hours before the production of sepsis. Sepsis in mice was induced by caecal ligation and puncture. Aortic rings were mounted in organ bath. Sepsis significantly ($P < 0.001$) decreased the mean arterial pressure (38.14 ± 1.32 mmHg, n = 6) in comparison to sham-operated (SO) controls (88.59 ± 5.07 mmHg, n = 6). Daidzein pre-treatment significantly ($P < 0.001$) attenuated the sepsis-induced fall in mean arterial pressure (68.38 ± 3.72 mmHg, n = 7). SO mice exhibited concentration-dependent contractions to noradrenaline (1 nM-10 μM) in mouse aorta. The E_{max} and pD_2 values were 0.44 ± 0.03 g and 7.28 ± 0.10 (n = 7), respectively. Sepsis significantly ($P < 0.01$) decreased the contractile responses to noradrenaline (E_{max} 0.16 ± 0.04 g and pD_2 7.02 ± 0.31; n = 6). Daidzein was unable to preserve contractions to noradrenaline. Endothelium intact aortic rings from SO mice developed sustained contraction 0.53 ± 0.02 g, n = 25 (rings) to K$^+$ (80 mM). Sepsis significantly ($P < 0.05$) decreased the contraction to 0.40 ± 0.02 g, n = 27 (rings). Daidzein pre-treatment in sepsis significantly ($P < 0.01$) restored KCl-induced contractions to 0.55 ± 0.02 g, n = 24 (rings), which were comparable to SO controls. Phenylephrine (1 μM)-pre-contracted endothelium intact rings from aorta of SO mice exhibited concentration-dependent relaxations to ACh (1 nM-10 μM) with pD_2 and E_{max} of 7.47 ± 0.07 and $92.75 \pm 2.64\%$ (n = 6), respectively. Sepsis significantly ($P < 0.05$) reduced the concentration-dependent relaxations to ACh with pD_2 and E_{max} of 7.33 ± 0.06 and $67.78 \pm 2.79\%$ (n = 13 rings), respectively. Daidzein treatment in septic mice not only completely restored the maximal relaxation (E_{max} $95.17 \pm 0.76\%$; $P < 0.001$) to ACh but also significantly increased the potency of ACh ($pD2$ 7.77 ± 0.10; n = 6) in comparison to septic animals.

CONCLUSIONS: Daidzein pre-treatment in septic mice prevents hypotension and vascular dysfunctions induced by CLP.

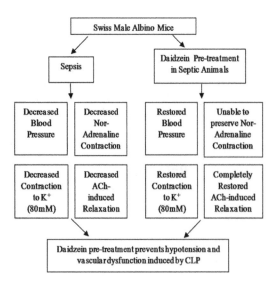

Adrenal-Specific G Protein-Coupled Receptor Kinase (GRK)-2 Deficiency Reduces Circulating Catecholamine Levels and Improves Cardiac Function after Myocardial Infarction

Anastasios Lymperopoulos[1], Giuseppe Rengo[2], Erhe Gao[3], Steven N. Ebert[4], Gerald W. Dorn[5] and Walter J. Koch[3]

[1]*Nova Southeastern University, Fort Lauderdale, Florida, United States;* [2]*University "Federico II", Naples, , Italy;* [3]*Temple University, Philadelphia, Pennsylvania, United States;* [4]*University of Central Florida, Orlando, Florida, United States;* [5]*Washington University, St. Louis, Missouri, United States*

BACKGROUND: Chronic heart failure (HF) is characterized by sympathetic overactivity and enhanced circulating catecholamines (CAs), which significantly increase HF morbidity and mortality. Adrenal G protein-coupled receptor kinase-2 (GRK2) is up-regulated in chronic HF, leading to enhanced CA release via desensitization/downregulation of the chromaffin cell α_2-adrenergic receptors (α_2ARs) that normally inhibit CA secretion. Adrenal GRK2 blockade decreases circulating CAs and improves cardiac inotropic reserve and function.

OBJECTIVES: We have tested the effects of adrenal-targeted GRK2 gene deletion on circulating CA levels after myocardial infarction (MI) and, consequently, also on cardiac function in vivo.

METHODS AND RESULTS: To specifically delete GRK2 in the chromaffin cells of the adrenal gland, we crossed PNMTCre mice, expressing Cre recombinase under the chromaffin cell-specific phenylethanolamine N-methyl transferase (PNMT) gene promoter, with *floxed* GRK2 mice. After confirming a significant ($\sim$50 %) reduction of adrenal GRK2 mRNA and protein levels, the PNMT-driven GRK2 knockout (KO) offspring underwent myocardial infarction (MI) to induce HF. At 4 weeks post-MI, plasma levels of both norepinephrine and epinephrine were significantly reduced in PNMT-driven GRK2 KO, compared to control mice (734 ± 15 vs. 900 ± 35 pg/ml for norepinephrine, 434 ± 11 vs. 534 ± 13 pg/ml for epinephrine, respectively, $p < 0.05$, $n = 8$), suggesting markedly reduced post-MI sympathetic activation. This translated into improved cardiac function (ejection fraction post-MI: $41 \pm 1\%$ vs. $29 \pm 3.5\%$, for KO and control mice, respectively, $p < 0.05$, $n = 8$) and dimensions as well as amelioration of abnormal cardiac βAR signaling at 4 weeks after the MI.

CONCLUSIONS: Adrenal chromaffin cell-targeted GRK2 knockout decreases circulating CAs by restoring the sympatholytic function of adrenal α_2ARs, which, in turn, results in improved cardiac function and β-adrenergic reserve in post-MI HF.

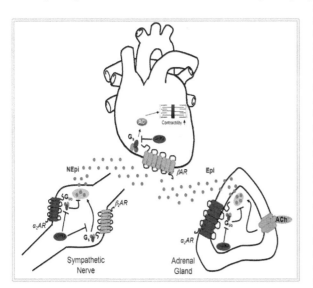

Serotonin and Catecholamines In Regulation of Myocardium Functions in Children with Congenital Heart Diseases

Razina Nigmatullina, Alvar Mustafin, Aidar Nigmatullin and Diana Bilalova

Department of Physiology, Kazan State Medical University, Kazan, Tatarstan, Russia

BACKGROUND: Serotonin (*5-HT*) has morphogenetic actions during embryonic period (Ugrumov, 1994; Nebigil, 2000) and it is involved in pathogenesis of cardiovascular diseases (Herve, 1995). The influence of *5-HT* on myocardial contraction in tissue from children with congenital heart disease (CHD) has not previously been investigated.

OBJECTIVES: To measured [5-HT], epinephrine [E] and norepinephrine concentrations [NE] in blood, the contraction force of children's atrial strips and expression of *5-HT2B* and *5-HT4* receptors (R) and the membrane serotonin transporter (*SERT*) in human right atria strips by immunohistochemistry.

METHODS AND RESULTS: Objects were 99 children with CHD at the age from 2 months till 17 years. [5-HT] in platelets at children with CHD does not change, but [5-HT] in plasma increases in 3 times with the years. However, [5-HT] in 4 times above, than [E] in children with CHD. Plasma of [5-HIAA] at children with CHD, having pulmonary arterial hypertension (PAH) are raised, authentically correlating with a degree of a PAH, and can serve as a marker for an estimation of gravity and efficiency of therapy at this patients. [5-HT] in plasma at children with PAH in 2 times exceeds control indicators. Relation [5-HIAA]/[5-HT] in plasma, is maximum at children with PAH. [5-HT] in platelets at children with PAH on 23% more low, than in the control. For the first time, the presence of *5HT-2BR* and *5-HT4R*, and *SERT*, in the myocardium of children having CHD has been shown. During development the role of *5HT2R* in atrium contraction is decreased, but the *5HT4R* is active.

CONCLUSIONS: Serotoninergic system actively participates in the pathogenesis of PAH in children with CHD.

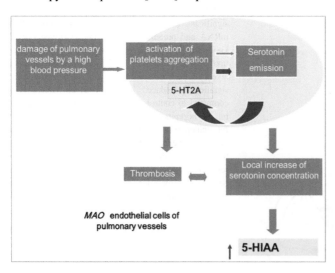

"Genetic and Acquired Determinants of Postural Tachycardia Syndrome"

Satish R Raj, MD MSCI

Autonomic Dysfunction Center, Departments of Medicine & Pharmacology, Vanderbilt University School of Medicine

Postural Tachycardia Syndrome (POTS) is likely the most common disorder seen in specialty Autonomic Clinics. It is a chronic disorder associated with excessive orthostatic tachycardia, in the absence of orthostatic hypotension, and a myriad of symptoms. POTS is not a single disease. Rather, it is best viewed as a "disorder" or a syndrome in which excessive orthostatic tachycardia can be a final common pathway of many underlying pathophysiological processes. In many cases, these patients are hyperadrenergic as measured by standing norepinephrine levels, and this likely contributes to the elevated heart rates.

This talk reviewed some of the variable pathophysiology that can develop into a POTS presentation. These can include "Mast Cell Activation Disorder", "Neuropathic POTS", and problems with hypovolemia and blood volume regulation and a low stroke volume state, all of which can lead to a secondary hyperadrenergic state. "Norepinephrine Transporter (NET) Deficiency", caused by a genetic loss of function mutation, has been identified in one kindred. This mechanism also had implications for the use of some common medications that act through NET inhibition. Recently, investigators from the Baker Institute have found that NET protein expression can vary among individuals and that there may be relatively less NET expression in patients with POTS. The identification of these individual pathophysiological mechanisms will hopefully serve as a starting point to the development of more rationale, pathophysiology-targeted treatments.

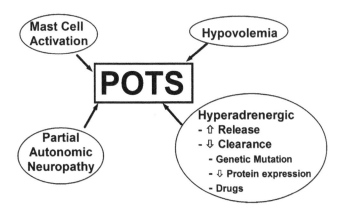

Neurotransmitter dopamine regulates tumor growth

Sujit Basu

[1]*Department of Pathology, Ohio State University, Columbus, Ohio, United States*

BACKGROUND: Dopamine is a catecholamine neurotransmitter and its roles in the pathogenesis of Parkinson's disease and schizophrenia are well established. However, recent reports indicate a unique correlation between this neurotransmitter and cancer. Angiogenesis or formation of new blood vessels is a critical process limiting growth of tumors. Vascular endothelial growth factor (VEGF) is one of the principal growth factors acting as a pro-angiogenic molecule in majority of malignant tumors.

RESULTS: We for the first time have demonstrated a novel link between the nervous system and tumor neovascularization. Dopamine (DA) by acting through its D_2 receptors can inhibit angiogenesis (microvessel permeability, adult endothelial cell migration and proliferation) and vasculogenesis (mobilization of bone-marrow derived endothelial progenitor cells to tumor microenvironment) by suppressing VEGF receptor 2 phosphorylation (VEGFR-2) in tumor endothelial cells (TEC) and endothelial progenitor cells (EPC) and thereby, inhibit growth of several malignant tumors in pre-clinical animal models. Importantly, DA also by acting through its D_2 receptors can normalize abnormal tumor blood vessels by up-regulating angiopoietin (Ang 1) expression in pericytes and Kruppel-like factor-2 (KLF2) expression in TEC leading to the delivery of higher concentration of conventional anticancer drugs to the tumor tissues.

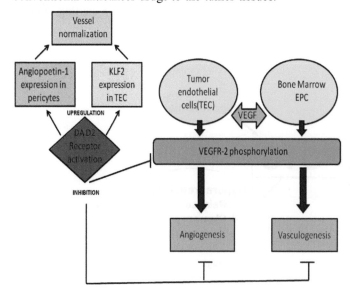

CONCLUSIONS: DA controls the functions of tumor blood vessels by regulating the functions of TEC and EPC, which in turn not only inhibit tumor growth, but can also increase the efficacy of conventional anti-cancer treatment. These results thus can be translated from the bench to the clinics for the treatment of cancer patients as dopamine D_2 receptor agonists are already in clinical use for the treatment of other diseases.

Extra Virgin Olive Oil Increases Uncoupling Protein 1 Content in Brown Adipose Tissue and Enhances Noradrenaline and Adrenaline Secretions in Rats

**Yuriko Oi-kano[1], Teruo Kawada[2], Tatsuo Watanabe[3], Fumihiro Koyama[4],
Kenichi Watanabe[4] and Kazuo Iwai[1]**

[1]*Kobe Women's University, Kobe, Hyogo, Japan;* [2]*Kyoto University, Uji, Kyoto, Japan;* [3]*University of Shizuoka,
Shizuoka, Shizuoka, Japan;* [4]*J-oil Mills, Yokohama, Tokyo, Japan*

BACKGROUND: Olive oil is an integral ingredient of the Mediteranean diet. In particular, extra virgin olive oil (EVOO) contains various minor components (phenolic compounds) that produce a particular aroma and taste.

OBJECTIVES: The effects of EVOO on triglyceride metabolism were investigated by measuring the degree of thermogenesis in interscapular brown adipose tissue (IBAT), and the rates of noradrenaline and adrenaline secretions in rats, both *in vivo* (**Experiment1**) and *in situ* (**Experiment 2**). This study was carried out to identify the constituents of EVOO that are effective in enhancing triglyceride catabolism.

METHODS AND RESULTS: In Experiment1, rats were given an isoenergetic high-fat diet containing corn oil, refined-olive oil or EVOO for 28 days. The phenolic contents in refined-olive oil and EVOO were 0 and 141 mg (oleuropein aglycone 104 mg)/kg respectively. After 28 days of feeding, the final body weight, perirenal adipose tissue, and epididymal fat pad and plasma triglyceride concentrations were the lowest in rats fed EVOO diet. The content of UCP1 in IBAT and the rates of urinary noradrenaline and adrenaline excretions were the highest in the rats fed the EVOO diet. In Experiment 2, the effects of the extract of phenolic fraction from EVOO and the three compounds having excellent characteristics as components of EVOO, oleuropein, oleuropein aglycone and hydroxytyrosol were evaluated. The intravenous administration of the extract of the phenolic fraction from EVOO significantly increased plasma noradrenaline and adrenaline concentrations, whereas that of hydroxytyrosol had no effect. The intravenous administration of oleuropein and oleuropein aglycone significantly increased plasma noradrenaline and adrenaline concentrations. Furthermore, oleuropein aglycone induced the secretions of noradrenaline and adrenaline about ten fold more potently than oleuropein.

CONCLUSIONS: These results suggest that the phenolic compounds, oleuropein and oleuropein aglycone in EVOO enhance thermogenesis by increasing the UCP1 content in IBAT and noradrenaline and adrenaline secretions in rats.

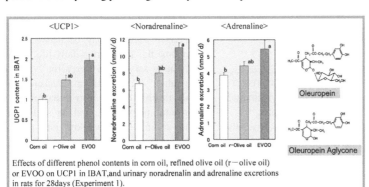

Effects of different phenol contents in corn oil, refined olive oil (r—olive oil) or EVOO on UCP1 in IBAT,and urinary noradrenalin and adrenaline excretions in rats for 28days (Experiment 1).

Neurovisceral Integration and Regulation of Peripheral Indices of ANS Function

Julian F. Thayer
The Ohio State University

The intimate connection between the brain and the visceral organs was enunciated by Claude Bernard over 150 years ago. In our neurovisceral integration model we have tried to build on this pioneering work. In the present paper we further elaborate our model and update it with recent results. Specifically, we performed a meta-analysis of recent neuroimaging studies on the relationship between heart rate variability and regional cerebral blood flow. We identified a number of regions, including the amygdala and ventromedial prefrontal cortex, in which significant associations across studies were found. In addition we provide evidence that more dorsal regions of the anterior cingulate may be associated with sympathetic outflow whereas more ventral regions may be associated with more parasympathetic outflow. These results are consistent with the notion of a medial prefrontal-brainstem visceromotor circuit involved in the regulation of the peripheral organs of the autonomic nervous system. We further propose that the default response to uncertainty is the threat response and may be related to the well known negativity bias. Heart rate variability may provide an index of how strongly 'top −down' appraisals, mediated by cortical-subcortical pathways, shape brainstem activity and autonomic responses in the body. If the default response to uncertainty is the threat response, as we propose here, contextual information represented in 'appraisal' systems may be necessary to overcome this bias during daily life. Thus, HRV may serve as a proxy for 'vertical integration' of the brain mechanisms that guide flexible control over behavior with peripheral physiology, and as such provides an important window into understanding stress and health.

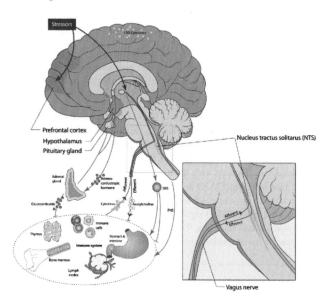

FIGURE 1

Schematic illustration of the connections between the brain and the immune system. Signalling between the immune system and the central nervous system (CNS) through systemic routes, the vagus nerve, the hypothalamic−pituitary−adrenal (HPA) axis, the sympathetic nervous system (SNS), and the peripheral nervous system (PNS) are shown. The afferent and efferent vagus meet at the nucleus of the solitary tract (shown in the insert).

Catecholamines in Integrative Function

Lee Eiden and David Goldstein

As the title of this theme implies, there was much overlap with many of the previous themes. Indeed, one might presume that it could have been completely subsumed within the previous more 'specific' themes. However, the integrative theme, organized by Esther Sabban and George Chrousos, featured and articulated progress and problems in critical areas of translational importance that in fact deserved special emphasis.

The chief topic among these was post-traumatic stress disorder (PTSD). The role of catecholamines, and neuropeptides such as NPY co-released with catecholamines in PTSD has been clearly recognized for more than 20 years, and in fact an entire book devoted to the role of catecholamines in PTSD was published 18 years ago (M. Michele Murburg, Ed., "Catecholamine Function in PTSD: Emerging Concepts", 1994). Nevertheless, and for a variety of reasons including advances in clinical chemistry, patient diagnosis, and unfortunately greatly increased patient numbers, progress in understanding the role of catecholamines and neuropeptides in PTSD has been slow until quite recently. Contributions by D. Baker and colleagues establishing NPY as a PTSD biomarker, M. Ruskind and colleagues in connecting locus coeruleus function and dysfunction to PTSD, and

Pervanidou and Chrousos in linking childhood PTSD to both the SNS and HPA axes have been important milestones, all featured in the second session of this Theme[1].

A presentation by the group of Sabban et al., on the role of NPY in attenuating PTSD-like symptoms in a rat model of PTSD (single severe, prolonged stressor administration) provided a seque between PTSD and mechanisms and consequences of chronic stress, a session commencing with a perspective offered by George Chrousos on the stress response as a homeostatic mechanism affording pathophysiological consequences only under certain conditions, and perhaps also only for certain individuals. Susceptibility, versus resilience, to dysfunction and/or disease following chronic stress may ultimately be revealed by a systems biological profile of the stressed individual. This will require a systems biological approach that includes immunological, hormonal, and metabolic aspects of response to stress, to identify disturbance in the 'return to homeostasis' in each of these systems, for a given stressor or series of stressful life experiences. David Morilak's thought-provoking analysis of the role of prefrontal cortical noradrenergic systems in modulating cognitive flexibility in response to both acute and chronic stress was a striking complement to Chrousos' presentation, particularly in offering a rationale for cognitive behavioral therapy for depression associated with chronic stress.

The relationship between obesity, metabolic regulation, depression and stress was explored through presentations by Richard Kvetnansky, who outlined a careful series of experiments in experimental animals revealing that adipocytes contain the components for catecholamine synthesis, storage and release; and by Julio Licinio, the chair of the session, who called attention to the therapeutic possibilities inherent in identifying the signaling molecules associated with adiposity and depression, in order to curb both.

Theme I concluded with an important session on catecholamines and hypoxic stress. Speakers in this session, chaired by Dona Wong, outlined the molecular connections between HIF signaling to catecholamine biosynthetic gene expression and its modulation by cytokines under conditions of inflammation, changes in HIF1 and HIF2 expression in pheochromocytoma, interactions between stress and hypoxia at the level of the adrenal medulla, and the role of reactive oxygen species in NPY-mediated modulation of catecholamine biosynthesis. Collectively, these presentations revealed how catecholamines play a critical role in both the mechanisms of detection and homeostatic response to hypoxia, and, importantly, in pathophysiological (allostatic) consequences of the hypoxic response.

---

[1]Note-session numbers correspond to the six session time slots of the XICS meeting. Since co-sponsored sessions are here associated with the 'lead theme', some Themes have less than six sessions.

Neuropeptide Y Infusion to Rats Attenuates Development of PTSD-Like Symptoms to Traumatic Stress

Esther L. Sabban, Lidia I. Serova, Andrej Tillinger, Lishay G. Alaluf and Marcela Laukova
Department of Biochemistry and Molecular Biology, New York Medical College, Valhalla, New York, 1059

BACKGROUND: PTSD is a debilitating psychiatric disorder induced by exposure to a severe traumatic stress and is associated with enhanced noradrenergic activity and dysregulation of the HPA axis. The treatment of PTSD is extremely challenging, and improved therapies and prevention techniques are greatly needed. Neuropeptide Y (NPY) is implicated in the etiology and pathophysiology of mood disorders as well as in stress regulation, coping and resilience to PTSD. However, due to its hypertensive effects, peripheral administration of NPY is undesirable.

OBJECTIVES: To determine if direct delivery of NPY to the CNS, by intranasal (IN) infusion, bypassing the blood—brain barrier, will prevent or reduce harmful effects of traumatic stress and reduce the development of PTSD like symptoms.

METHODS AND RESULTS: NPY or vehicle were administered IN to rats 30 min pre- or immediately post-exposure to single prolonged stress (SPS) model of PTSD. Animals pretreated with IN NPY, compared to vehicle, spent significantly less time immobile during the first and second 5 min intervals of the forced swim part of the SPS. Thirty min after SPS, TH and DBH mRNAs were elevated in locus coeruleus (LC) of the vehicle treated rats compared to unstressed controls. IN NPY prevented this SPS elicited increase in TH and DBH mRNA levels. Moreover, the SPS triggered rise of plasma ACTH and corticosterone were attenuated by IN NPY pretreatment. Seven days after SPS, animals pretreated with IN vehicle, but not IN NPY, still had plasma ACTH and corticosterone levels significantly above unstressed controls. Glucocorticoid receptor (GR) protein levels were determined by western blots in several brain areas implicated in features of PTSD. There was elevated GR expression in the hippocampus and LC, but not in the amygdala or prefrontal cortex. Pretreatment with NPY prevented the SPS-elicited elevation of GR. One or two weeks after SPS, the vehicle treated animals developed several PTSD-like behavioral symptoms: increasing anxiety-, depressive-like behaviors and hyper arousal as evaluated by Elevated Plus Maze, Forced Swim Test and Acoustic Startle Response. Administration of NPY, either 30 min pre- or immediately post- SPS, elicited long-term improvements in anxiety-, depressive-like behavior and acoustic startle responses.

CONCLUSIONS: IN NPY attenuated short-term as well as long-term neuroendocrine and behavioral responses to SPS. The results provide proof of principle that rapid delivery of NPY to the brain by single IN infusion before or shortly after SPS has a pronounced resilient effect ameliorating development of PTSD-like symptoms.This non-invasive procedure has therapeutic potential to prevent or reduce harmful effects of traumatic stress.

Supported by grant DM102881 from the US Army, Department of Defense Medical Research and Development Program.

PTSD and the Locus Coeeurleus

Patricia Szot, Elaine Peskind and Murray Raskind

MIRECC, VA Puget Sound Health Care System, USA

Increased CNS noradrenergic activity contributes to the pathophysiology of posttraumatic stress disorder (PTSD). Excessive responsiveness of the postsynaptic alpha-1 adrenoreceptor (AR) to norepinephrine is an attractive candidate mechanism. Alpha-1 AR agonists disrupt REM sleep, increase the release of the anxiogenic neuropeptide corticotrophin releasing factor, and act on prefrontal cortex to favor "fight or flight" cognition. The most compelling evidence supporting increased alpha-1 AR activity in PTSD comes from randomized controlled trials (RCTs) of the alpha-1 AR antagonist prazosin in PTSD. In three RCTs in combat trauma PTSD, prazosin was effective for trauma nightmares, sleep disruption and global clinical function without producing sedation. In two of these studies prazosin also was effective for overall PTSD symptomatology. In a placebo controlled trial in civilian PTSD, prazosin increased total sleep time by 94 minutes, reduced trauma nightmares and increased and normalized REM sleep. Recent studies suggest that normal REM sleep (and normal dreaming) is important for reducing or eliminating excessive emotional tone from memories. Prazosin is also effective for agitation/aggression in Alzheimer's disease (AD) and postmortem studies demonstrate up-regulation of alpha-1 AR in AD frontal cortex and hippocampus. What is unclear at the present time is why the alpha-1 AR response is enhanced in PTSD and AD. Preliminary work in our laboratory has demonstrated a significant increase in alpha-1 AR binding sites in the frontal cortex as well as bed nucleus of the stria terminalis 3 days after the bilateral administration of 6 hydroxydopamine to lesion locus coeruleus neurons. However, 3 weeks after 6-hydroxydopamine administration alpha-1 AR binding sites are not elevated in the CNS. Work is in progress to determine the cause of the elevated alpha-1 AR binding sites.

Systems Biology and the Stress Response: From Pythagoras and the Epicureans to Modern Medicine

George Chrousos

Medical School of the University of Athens, Greece

"Stress" is defined as the disturbance in the "dynamic balance" or "homeostasis" of a complex system, such as the human organism or society, "stressors" as the forces that produce this disturbance, and "adaptive response" as the forces from within the complex system that strive to return homeostasis to normal. The adaptive response is subserved by a specialized system in the human brain and body, the "Stress System", which is activated when a stressor of any kind exceeds a certain threshold. In itself, stress is neutral or even potentially beneficial. In contrast, the chronic dynamic state in which the adaptive response fails to fully reestablish homeostasis during stress, which I call "dyshomeostasis" or, more correctly, "cacostasis", may have detrimental effects. The crucial beneficial hormones that are activated to reestablish homeostasis, the "homeostatic or stress mediators", including adrenaline, noradrenaline, cortisol and the inflammatory mediator interleukin-6, may also, paradoxically, be responsible for the damage sustained in cacostasis. These mediators may impair cellular physiology and metabolism, increase inflammatory responses and stimulate oxidative functions. These changes can ultimately accelerate aging, cause obesity, metabolic problems and osteoporosis, promote atherosclerosis and cardiovascular disease and increase the chances of infection or cancer. The pre-Socratic philosophers Pythagoras and Alcmaeon, respectively, used the terms "harmony" and "isonomia" to express the dynamic balance or homeostasis of life, while the Hippocratics equated this harmony with health and disharmony with disease. Both the Stoics and the Epicurians, philosophic schools that concentrated on the study of stress and its management, considered the attainment of "ataraxia", or imperturbability of the mind to stressors, as the ultimate goal of life, while Epicurus himself spoke of "eustatheia" -or "eustasis", if we extrapolate from homeostasis-, the serene emotional state of a harmonious balance in a human being. The not uncommon Greek first name "Eustathios " is a remnant of that era. Based on these ancient seminal ideas, we suggested that appropriate responsiveness of the human stress system to stressors is a crucial prerequisite for a sense of wellbeing, adequate performance of tasks, and positive social interactions, and hence for the survival of the self and the species. By contrast, inappropriate, over- or under- responsiveness of the stress system through its hormonal and inflammatory mediators may impair growth and development, and may account for the many chronic behavioral, endocrine, metabolic, and allergic/autoimmune disorders that plague contemporary humanity. The development and severity of these conditions primarily depend on the genetic vulnerability of the individual, the exposure to adverse environmental- including psychosocial and economic- factors and the timing and duration of the stressful event(s). Our first model disorders of stress system over- and under-responsiveness were, respectively, melancholic depression, which we showed to be a disease characterized by chronic hyperactivation of the stress system, explaining its behavioral and somatic manifestations, and its mirror image, atypical depression, characterized by chronically decreased activation of this system, also explaining its clinical manifestations through a similar but alternative pathway. This realization allowed us to extrapolate our thoughts to the larger nosology of the human population, identifying many other such states. Starting from animal models, and with some surprise, we realized that hyporesponsivity of the stress system to inflammatory signals and resistance of immune tissues to cortisol characterize several inflammatory disorders, such as rheumatoid arthritis and septic shock, phenomena that we elucidated at the molecular level. Similarly, we identified other human states characterized by hypoactivity of the stress response, including seasonal depression, the chronic fatigue and fibromyalgia syndromes, postpartum blues/depression and the late luteal phase dysphoric syndrome disorder. We suggested that the genetic vulnerabilities of our species leading to the contemporary "chronic non-communicable diseases", which include obesity, the chronic pain and fatigue syndromes, depression, hypertension, the metabolic syndrome and cardiovascular diseases, resulted from selective pressures of evolutionary stressors upon our genome during our evolution to modern human beings. To these genetic vulnerability changes, stress in prenatal and early life adds the so-called epigenetic ones. Prenatal life, infancy, childhood and adolescence are critical periods characterized by great plasticity and, hence, increased vulnerability to stressors. During these critical periods of life, stress mediators exert major organizational effects on the neural circuits of the brain and epigenetic effects throughout the brain and body, i.e., effects that are prolonged or permanent and include chemical changes in our DNA and chromatin. Such epigenetic changes are environmentally acquired, and as such, are fully preventable. Well beyond depression, currently affecting approximately 20% of the adult population and predicted by the World Health Organization to soon be the number 2 morbidity factor in the world, we suggested that chronic common, every day stress and the resultant cacostasis are also a major factor for all cause morbidity and mortality in today's societies, encompassing all the chronic non-communicable diseases and, to a lesser extent, certain infections and cancer. At this time, and despite the stupendous advances in hygiene and medicine that have prolonged our lives, we believe that over 50 percent of the human disease burden is due to chronic socioeconomic distress, a result -to a great extent- of the ever increasing complexity and alienation in the modern world. We should note that the effects of prolonged distress go well beyond the causation of the chronic non-communicable diseases. Prolonged human cacostasis disturbs normal growth and development in children, causes emotional deprivation, upsets sleep, destroys family and social life, increases criminal behavior, multiplies violent accidents, and stimulates the abuse of and dependence on substances. Chronic distress also accelerates aging, reflected in decreasing chromosomal telomere length, and increases frailty, curtailing healthy life expectancy. Finally, the health care related costs of prolonged distress are enormous, probably accounting for a significant proportion of the total expenditures. The dysphoria that most frequently accompanies cacostasis and the "happiness" of equanimity and sense of wellbeing are mutually inhibitory of each other. We now understand the neurochemical mechanisms of this crucial relation and can employ this knowledge in our quest for happiness and good health. We actually can, with the power of our will, lead ourselves to a dynamic state of homeostasis that is at a level higher than what would have been expected from our genetic and epigenetic constitution and our environment, a condition that I call "hyperstasis". The latter state allows the full experience and appreciation of the gift of life, extends its duration and influences propitiously those around us. This state gives materialism its true proportions and promotes virtue, spirituality and a better Society.

Through our work, we established the common biological pathways connecting a large array of ostensibly disparate psychological and somatic disorders, including depression, anxiety, obesity, hypertension, metabolic syndrome, diabetes, allergic and autoimmune inflammatory disorders, sleep disturbances, such as insomnia and sleep apnea, hypofertility, and osteoporosis. We contributed to the understanding of how stress through its mediators, including components of the inflammatory reaction, causes premature aging, and promotes cardiovascular and neurovascular diseases. A common pathophysiology suggests that prevention and curative means have common bases. Today, the presence of stress and cacostasis in an individual can be evaluated and graded. There are rational and proven methods to prevent and ameliorate distress that start from changes in lifestyle (healthy diet, exercise, stable daily timing, adequate sleep), to cognitive and behavioral therapies, to the use of appropriate medications. Indeed, the currently available medications that control risk factors and prolong life function primarily by blocking the stress system and inflammatory mediators. The key issue is that nature is not destiny and wellbeing and hyperstasis are attainable. Granted that chronic distress in early life augments the risk of developing chronic behavioral and non-communicative disorders, preventing distress in pregnancy and the first 5 years of life or interrupting the vicious cycle of distress during this period is imperative and, in the long run, the most cost-effective approach. Interventions beyond the age of 5 may be quite useful, but one should note that major stress-related brain organizational and epigenetic damage has already occurred. At this point I should mention what L. Tolstoy had intuitively said: "From the child of five to myself is but a step. But from the newborn baby to the child of five is an appalling distance". Now, this is based on robust evidence and makes early interventions, starting with the education of prospective mothers, a must. In summary, to interrupt the vicious effects of stress in a society and its members, one should, first, eliminate or at least moderate the stressors and, second, one should improve the coping of individuals with such stressors - aka improve their resilience to stress-. Political actions can influence both strategies: Granted that stress in today's industrialized world is mostly anthropogenic, a well run country itself in homeostasis, in which people feel enfranchised, dignified and dealt with fairness and justice, is bound to have happier and healthier citizens. On the other hand, preventing early life stressors and their effects on the very young will eliminate development of risks for the later behavioral and chronic non-communicative disorders that plague our societies today. Finally, like many human endeavors, coping with stress is an eminently learnable process and the basis of the most effective psychological therapy employed today, cognitive behavioral therapy. It is the duty of a Society to ensure the wellbeing and happiness of its people with political strategies that prevent stress and enhance the ability of its citizens to cope.

Reactive Oxygen Species and Neuropeptide Y Signaling in Altered Catecholamine Synthesis during Intermittent Hypoxia

Ganesh K. Kumar, Gayatri Raghuraman and Nanduri R. Prabhakar

Institute for Integrative Physiology and Center for Systems Biology of Oxygen Sensing, Biological Sciences Division, Department of Medicine, University of Chicago, Chicago, IL 60637, USA

Humans encounter two forms of chronic hypoxia; continuous hypoxia (CH; experienced during high altitude sojourns) and intermittent hypoxia (IH; encountered under a variety of conditions including sleep disordered breathing manifested as recurrent apneas). Both CH and IH reduce oxygen availability; however, organisms adapt to the effects of chronic CH whereas IH adversely impacts autonomic functions leading to cardio-respiratory morbidities. Catecholamines (CA) play important roles not only in development and energy metabolism, but also in regulation of cardio-respiratory functions during hypoxia. They are expressed in the brainstem and adrenal medulla associated with the regulation of blood pressure. Studies were initiated in rodents and cell cultures to assess the effects of IH on CA synthesis and to unravel the underlying cellular mechanism(s). Analysis of the enzyme activity of tyrosine hydroxylase (TH), the rate-limiting enzyme in CA synthesis showed that IH increased TH activity and CA levels without any accompanying changes in TH protein expression. The IH-induced increase in TH activity i) is evident in the brainstem and adrenal medulla of rats as well as in PC12 cells, ii) seems to be dependent on the pattern of IH, iii) is associated with an increase in the generation of reactive oxygen species (ROS) and iv) is, in part, due to increased serine phosphorylation of TH. IH increased protein kinase activity whereas it decreased protein phosphatase activity which seems to contribute to the sustained elevation in serine phosphorylation of TH. In addition to CA, adrenal medulla also expresses neuropeptide Y (NPY) which is known to modulate TH activity via activation of G-protein coupled Y1 receptors (Y1R). Therefore, the role of NPY in IH-induced TH activation was examined. IH markedly elevated NPY levels, Y1R mRNA and protein expression, cAMP levels and protein kinase A (PKA) activity. Systemic administration of 4-phenyl-3-butenoic acid (PBA), an inhibitor of bioactive NPY synthesis, prevented IH-induced increases in NPY levels, TH activity, TH phosphorylation (at Ser-40) and CA levels. Furthermore, IH-induced increase in TH phosphorylation and activity and up-regulation of PKA activity were absent in Y1R-/-mice. Treatment with antioxidant not only abolished IH-induced changes in TH activity and phosphorylation but also prevented the increases in NPY and Y1R expression. Rats and mice exposed to 10 days of IH show elevated mean arterial pressure. Adrenal demedullation as well as treatment with either PBA or antioxidant prevented IH-induced hypertension. Notably, acute administration of Y1R antagonist transiently lowers MAP in IH-exposed rats and mice, and IH-induced hypertension was absent in Y1R-/-mice. The above findings, collectively, suggest that ROS-NPY-Y1R-PKA signaling mediates IH-induced increase in CA synthesis in the brainstem and adrenal medulla which may contribute in part to the hypertension caused by IH associated with recurrent apneas.

Supported by grants from HL-89616 and HL-90554.

Stress, HIF1a, and Adrenergic Function

Karyn M. Myers[1], Robert Claycomb[1], T.C. Tai[2], William A. CarlezonJr.[1], Richard Kvetnansky[3] and Dona Lee Wong[1]

[1]*Harvard Medical School and McLean Hospital, Belmont, MA, United States;* [2]*Northern Ontario School of Medicine, Laurentian University, Sudbury, ON, Canada;* [3]*Institute of Experimental Endocrinology, Slovak Academy of Sciences, Bratislava, Slovakia*

BACKGROUND: The stress hormone epinephrine (EPI) plays an important role in activating physiological and behavioral responses enabling organisms to overcome stress and restore homeostasis. EPI is also associated with stress-induced illness, cardiovascular disease, immune dysfunction, cancer and behavioral disorders. Little is known about mechanisms underlying the role of EPI in these illnesses. Immobilization (IMMO) stress in rats and hypoxic stress (5% oxygen) in rat-derived PC12 cells are being used to examine mechanisms by which stress regulates EPI via its biosynthesis by phenylethanolamine N-methyltransferase (PNMT). Two PNMT transcriptional activators, Egr-1 and Sp1, are critical for stress activation of the PNMT gene. Hypoxia inducible factor 1 alpha (HIF1a) may serve as a master "on-off" switch under stressful conditions to regulate these factors and thereby, PNMT and EPI expression. Post-traumatic stress disorder (PTSD) continues to escalate, particularly in U.S. veterans returning from Iraq and Afghanistan, with estimates of over 213,000 PTSD casualties from year 2000 to the present. PTSD patients exhibit memory extinction deficits and show elevated EPI with symptom relief after adrenergic antagonist treatment. We have developed a rat PTSD model using IMMO stress and are using that model to demonstrate that stress-induced elevation of PNMT and EPI alters adrenergic drive via central learning and memory circuitry to provide long-term adaptive/maladaptive changes to stress responsiveness, with maladaptive changes having the potential to evoke illness.

OBJECTIVES: To reveal mechanisms by which stress alters adrenergic function to develop therapeutics for stress-induced illness that permit early intervention to circumvent disease.

METHODS AND RESULTS: Male Sprague Dawley rats are subjected to IMMO stress for 30 or 120 min daily for 7 days and adrenal medulla collected. Alternatively, rats are trained in fear extinction using the rat fear potentiated startle paradigm, followed by IMMO as described, and retested for fear extinction subsequent to IMMO. Cell culture methods include transient transfection with PNMT promoter-luciferase reporter gene constructs and exposure to 5% oxygen. Alternatively, cells are exposed to hypoxia and/or drugs or siRNA. As appropriate, tissue and cells are analyzed for transcription factor and PNMT mRNA and protein using RT-PCR and western analysis. Animal and cell culture studies demonstrate that two PNMT transcriptional activators, Egr-1 and Sp1, are elevated by both IMMO (single or repeated) and hypoxic stress, with changes in mRNA and protein expression. HIF1a is also induced by both stressors. siRNA knock down of Egr-1, Sp1 and HIF1a in cell culture suggests that HIF1a does not directly activate PNMT but does so by inducing Egr-1 and Sp1. The latter is consistent with undetectable binding of HIF1a to a putative hypoxia response element in the PNMT promoter and induction of Egr-1 and Sp1 by over expression of HIF1a in the cells. In rats trained in fear extinction, IMMO stress interferes with extinction memory consolidation and/or retrieval in the rat fear-potentiated startle (FPS) paradigm with marked corresponding elevation of adrenal adrenergic indices.

CONCLUSIONS: Stress regulates EPI via activation of the PNMT gene via induction of Egr-1 and Sp1. HIF1a serves as the "on-off" switch for stress activation by genetic control of Egr-1 and Sp1. As demonstrated through the rat fear extinction paradigm and assessment of adrenergic indices in the adrenal medulla, repeated IMMO provides a rat PTSD model for examining mechanisms underlying the role of stress-induced EPI elevation in illness.

Characterization of Cerebrospinal Fluid (CSF) and Plasma NPY Levels in Normal Volunteers over a 24-h Timeframe

Dewleen G Baker[1,2,3], **Richard L Hauger**[1,2,3], **Tobias Moeller Bertram**[1,2,3], **Piyush M Patel**[2,3], **Donald A Barkauskas**[4], **Paul Clopton**[3], **Thomas D Geracioti Jr**[5], **Daniel T O'Connor**[2] and **Caroline M Nievergelt**[2,3]

[1]*VA Center of Excellence for Stress and Mental Health, San Diego, CA, United States;* [2]*University of California San Diego, San Diego, CA, United States;* [3]*VA San Diego Healthcare System, San Diego, CA, United States;* [4]*University of Southern California, Los Angeles, CA, United States;* [5]*University of Cincinnati, Ohio, CA, United States*

BACKGROUND: Neuropeptide (NPY) is abundant in mammals, where it contributes to diverse functions centrally and peripherally. Its wide brain distribution provides a plausible substrate for its relevance to stress. Despite an increasing interest in NPY as a moderator of stress in humans, the extent to which plasma and cerebrospinal fluid NPY concentrations are accurate reflections of each other is poorly understood.

OBJECTIVES: The objective of this study is to more thoroughly characterize NPY CSF/plasma concentration relationships.

METHODS AND RESULTS: Eleven healthy male civilian study volunteers participated in a 24-h serial CSF and plasma sampling study. All met study inclusion criteria based on physical examination, mental health (DSM-IV) interviews. At 8AM the morning after admission, a catheter was inserted via a 17-gauge Touhy needle into the L4-L5 lumbar space. Between 11AM on day one CSF (and plasma, from an indwelling venous catheter) were collected. All fluids were stored at $-80°$C until assay of (hourly) samples. As observed in prior studies, group mean (SE) CSF NPY (cNPY) levels [792.1 (7.80) pg/mL] were higher than plasma (pNPY) levels [220.0 (3.63) pg/mL]. Lagged cross-correlation (CFF) analysis showed no statistically significant cross-correlations between cNPY and pNPY at the $p > .05$ level [see Figure]. Average pNPY/cNPY concentration ratios ranged from .20 to .40 across study subjects. The pNPY/cNPY ratios appear to be individual specific and consistent across the 24-h time period. cNPY circadian components were not detectable owing to a large positive linear trend.

CONCLUSIONS: These findings suggest that interpretation of the physiological significance of plasma NPY concentrations in human NPY stress or resilience studies must account for the lack of correlation between plasma and CSF NPY concentrations.

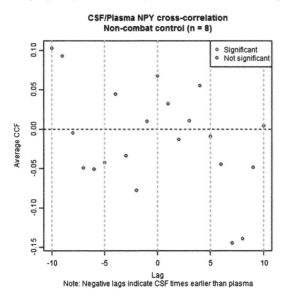

CSF/Plasma NPY cross-correlation Non-combat control (n = 8)

Legend: ○ Significant ○ Not significant

Y-axis: Average CCF; X-axis: Lag

Note: Negative lags indicate CSF times earlier than plasma

NPY: Low Cerebrospinal Fluid (CSF) Levels in Posttraumatic Stress Disorder in Comparison to Combat and Civilian, non-combat Control Subjects

Dewleen G Baker[1,2,3], Richard L Hauger[1,2,3], Tobias Moeller Bertram[1,2,3], Piyush M Patel[2,3], Donald A Barkauskas[4], Paul Clopton[3], Thomas D Geracioti[5], Daniel T O'Connor[2] and Caroline M Nievergelt[2,3]

[1]VA Center of Excellence for Stress and Mental Health, San Diego, CA, United States; [2]University of California San Diego, La Jolla, CA, United States; [3]VA San Diego Healthcare System, San Diego, CA, United States; [4]University of Southern California, Los Angeles, CA, United States; [5]University of Cincinnati, Cincinnati, OH, United States

BACKGROUND: The NPY system is associated with behavioral resilience to stress exposure in an animal model of Posttraumatic Stress Disorder (PTSD); its role in the humans with PTSD is being explored.

OBJECTIVES: The key objective of this 24-hour serial CSF study of NPY in PTSD was to replicate and expand upon a prior single time point PTSD study showing low CSF NPY, by evaluating basal 24-h NPY concentrations across three study groups, civilian volunteers and combatants of the Iraq and Afghanistan conflicts with and without PTSD.

METHODS AND RESULTS: Participants were 26 age-matched, males, 12 with PTSD, 14 healthy deployed and 11 civilians. After CSF catheter insertion, beginning at 11AM on study day one, CSF was collected every half hour for 24-h from an indwelling CSF catheter, as was plasma from a venous catheter. Fluids were stored at $-80°C$ until assay. Group demographic comparisons using FDR-adjusted p-values showed no statistically significant differences across study groups regard age or BMI. Using linear mixed-effect models, differences in NPY-CSF concentrations were statistically significant [p-value for existence of group effect 0.0234; the difference in levels between healthy civilian volunteers and PTSD subjects was statistically significant ($p = 0.012$) but deployed healthy subjects were not statistically significantly different from either of the other two groups]. Additionally, cNPY increased at an estimated rate of 5.5 pg/mL/hour ($p < 0.0001$) but tests for interaction showed no statistically significant differences in the linear trends among the three subject groups.

CONCLUSIONS: These findings suggest that NPY may be involved in behavioral resilience to stress in humans, thus may be a good target for interventions for prevention or early intervention.

Developmental Neuroendocrinology of Posttraumatic Stress Disorder in Children and Adolescents after Motor Vehicle Accidents

Panagiota Pervanidou and George P. Chrousos

Unit of Developmental and Behavioral Pediatrics, First Department of Pediatrics, University of Athens Medical School, "Aghia Sophia" Children's Hospital, Athens, Greece

Posttraumatic Stress Disorder (PTSD) is the most common trauma-related disorder in children. Similarly to adults, dysregulation of the hypothalamic−pituitary−adrenal (HPA) axis and the locus caeruleus/norepinephrine−sympathetic nervous system (LC/NE−SNS) is associated with the pathophysiology of the disorder. However, although catecholamines have been reported consistently high, diverse cortisol secretion patterns have been published. Genetic and epigenetic predisposing factors, age, developmental stage, early experiences (previous trauma), together with the nature of the trauma (chronic, repeated *vs.* single, acute trauma) and time since the trauma occurred influence the direction of HPA axis activity in PTSD.

Traumatic stress experiences in early life play a critical role for further traumatization and PTSD development. In addition, early life stress "programs" the developing brain to be more vulnerable and to react with more anxiety when confronted by new stressors. In children and adolescents after single, acute stressors such as motor vehicle accidents, we found high evening salivary cortisol and morning serum IL-6 concentrations to be predictive of persistent PTSD development 6 months later. In the longitudinal analysis, we demonstrated an initial elevation of evening salivary cortisol in the aftermath of the trauma, followed by a gradual normalization 6 months later, in parallel to a gradual elevation of −initially normal- norepinephrine over time, findings that represent the natural history of neuroendocrine changes in pediatric PTSD. This progressive divergence of evening salivary cortisol and morning plasma norepinephrine, from baseline to month 6, might be the underlying pathophysiologic mechanism responsible for PTSD development and maintenance over time. Posttraumatic symptoms may thus be attributed to cortisol dysregulation that fails to shut down the catecholaminergic response in limbic structures, such as *locus caeruleus* and the amygdala (1,2). Low cortisol and high norepinephrine is a common finding in adult PTSD and may reflect previous trauma, representing thus a late stage in the natural history of the disorder, as well as a biologic vulnerability factor for PTSD development in adulthood.

Proposed longitudinal divergence of cortisol and noradrenaline responsible for PTSD maintenance

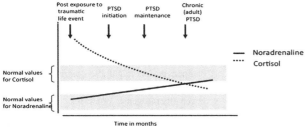

Pervanidou, J Neuroendocrinology, 2008, with permission

FIGURE

adapted from ref.[1]

References

1. Pervanidou P. Biology of post-traumatic stress disorder in childhood and adolescence. J Neuroendocrinol. 2008 May;20(5):632−8.

2. Pervanidou P, Chrousos GP. Neuroendocrinology of post-traumatic stress disorder. Prog Brain Res. 2010;182:149−60.

Repeated and Novel Stress-triggered Changes in Adipocyte Catecholamine System

B. Manz[1], K. Pacak[2], R. Kvetnansky, J. Ukropec, M. Laukova, B. Ukropcova and P. Vargovic

[1]*LDN, Nordhorn, Germany;* [2]*National Institute of Child Health & Human Development, NIH, Bethesda, USA; Inst. of Experim. Endocrinology SAS, Bratislava, Slovak Republic*

BACKGROUND: The sympathoadrenal system (SAS) is the main source of catecholamines (CAs) in adipose tissues where they play the key role in regulation of metabolism. We recently reported existence of an alternative CA producing system directly in adipocytes. Stress is an important factor in regulation of CA systems in mammals and activates SAS also in adipose tissue primarily in order to mobilize energy.

OBJECTIVES: The aim of this study was to investigate effect of a single and repeated exposure of rats to different stressors (emotional - immobilization, physical – cold) on CA production in adipocytes isolated from mesenteric adipose tissue, in order to shed some light to the processes of CA system adaptation to chronic influence of stressors and to responses of adapted animals to novel, heterotypic stressors specifically in the population of energy storage cells - adipocytes.

METHODS AND RESULTS: The presented data clearly demonstrate that NE *(figure1A)* but also EPI *(figure1B)* levels, and components of the CA biosynthetic apparatus - tyrosine hydroxylase (TH) *(figure1C mRNA, 1E protein)*, dopamine-β-hydroxylase (DBH), phenylethanolamine N-methyltransferase (PNMT, *(figure1D)* and VMAT1 transporter are present and dynamically modulated in adipocytes isolated from mesenteric adipose tissue of stressed rats. CAs were measured using 2-CAT or 3-CAT Research RIA kits (LDN, Nordhorn, Germany). Relative quantification of mRNA levels was done by real time RT-PCR. Cell specificity for CA synthesis in adipocytes was demonstrated by administration of alpha-methyl-p-tyrosine (TH antagonist) to freshly isolated adipocytes, as it largely reduced levels of detected CAs. Detailed analysis of dynamic responses of NE, EPI, TH mRNA, PNMT mRNA, and their protein levels to single and repeated immobilization stress *(time course 10, 30, 120 min)* clearly showed that repeated *vs.* acute stress leads to significantly exaggerated response of CA biosynthesis in isolated mesenteric adipocytes. However, animals adapted to a long-term cold exposure *(28 days, 4°C)* did not show responses similar to those found in adipocytes after a single IMO.

CONCLUSIONS: Our data indicate that adipocytes accommodate gene machinery able to synthesize NE and EPI, and that this process is highly activated under repeated stress. Thus, adipocyte CAs might also contribute to regulation of stress-related catabolic (lipolysis) or even thermogenic processes. However, the physiological function of CAs produced endogenou-sly in adipocytes has not been appointed yet.

Supported by APVV-0148−06 & APVV-0088-10; VEGA2/0188/09 & 2/0036/11.

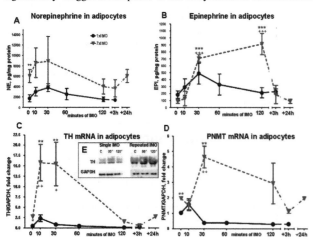

Jekyll and Hyde: Noradrenergic Modulation in Acute and Chronic Stress

David A. Morilak, Brian Bingham, Jennifer J. Donegan, Milena Girotti and Julianne D. Jett

Department of Pharmacology and Center for Biomedical Neuroscience; University of Texas Health Science Center at San Antonio, San Antonio TX, USA

BACKGROUND: Noradrenergic facilitation enhances adaptive behavioral and physiological responses to acute stress. But chronic stress is a risk factor for many psychiatric disorders, such as depression. Impaired executive function is a component of such disorders, specifically cognitive inflexibility (e.g., perseveration, cognitive-emotional biases). Cognitive flexibility, or the ability to modify established thoughts and behaviors in response to changes in the environment, requires the functional integrity of the medial prefrontal cortex (mPFC), which is hypoactive in depressed individuals.

OBJECTIVES: Investigate the acutely adaptive and chronically maladaptive modulatory effects of NE on cognitive flexibility in the mPFC.

METHODS AND RESULTS: We measured cognitive flexibility in rats using the Attentional Set-Shifting Test (AST). Chronic unpredictable stress (CUS) impairs cognitive set-shifting, and the CUS-induced deficit is prevented by chronic NE reuptake blockade. Alpha1 adrenergic receptors in mPFC facilitate set-shifting, and contribute to the effective antidepressant response. However, our most recent data indicate that repeated elicitation of noradrenergic facilitation in mPFC also contributes to the detrimental effects of chronic stress, as blockade of adrenergic receptors in mPFC during CUS protects against the cognitive flexibility deficit. Thus the next step will be to investigate the target of beneficial noradrenergic facilitation acutely, and of the detrimental effects of repeated noradrenergic activation chronically.

CONCLUSIONS: These results raise a paradox: how can elevating NE by chronic stress be bad, whereas elevating NE by chronic reuptake blockade is good? We hypothesize that the interaction of NE with other signaling pathways that are activated convergently during stress contributes to the detrimental effect, and we are currently investigating changes in cytokine and glutamate signaling in mPFC. We are also beginning to address mechanisms of novel therapeutic approaches, e.g., cognitive training to test the hypothesis that cognitive behavioral therapy engages the circuitry in PFC that is affected in depression, thereby improving function.

Supported by NIMH (MH053851 and MH072672)

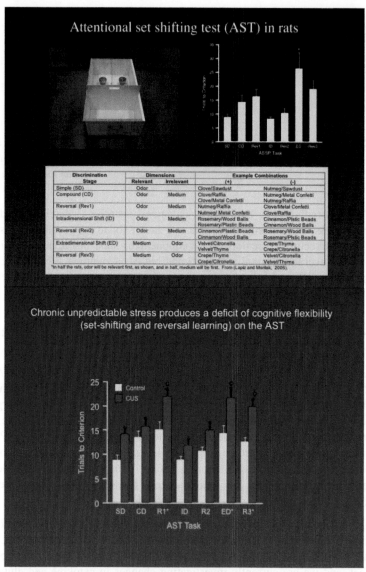

Bottom figure reproduced and adapted from: Bondi et al., 2008. Chronic unpredictable stress induces a cognitive deficit and anxiety-like behavior in rats that is prevented by chronic antidepressant drug treatment. *Neuropsychopharmacology*, 33: 320-331. Nature Publishing Group.

Early Life Stress Increases Nucleus Accumbens Dopamine Signaling

Jordan T. Yorgason, Jeffrey L. Weiner and Sara R. Jones

Wake Forest Health Sciences, Winston-Salem, NC, United States

BACKGROUND: Social isolation (SI) rearing in rats is a model of early life stress that results in disrupted behaviors modeled for depression, anxiety, schizophrenia, and drug addiction. Since DA modulates these disorders, it is notable that SI rearing results in DA neurochemical changes including increased DAergic firing, tissue levels, and greater responsivity to psychostimulants. Although previous studies suggest SI increases DA release, to date no study has measured rapid phasic DA signaling, uptake and autoreceptor activity in isolates.

OBJECTIVES: The current study uses voltammetric techniques to characterize the effects of SI rearing on DA terminal function, including release, uptake and autoreceptor activity.

METHODS AND RESULTS: Male Long-Evans rats (postnatal day 28) were placed in either SI (1 rat/cage) or GH (4 rats/cage) conditions for six weeks. Brains were then sectioned (400 μm), and NAc slices were perfused with oxygenated aCSF (32°C). Electrically evoked (1 pulse, 300 μA, 4 ms, monophasic) DA overflow was detected using cyclic voltammetry (-0.4 V to 1.2 V; 10 Hz; 400 V/s). Baseline data were analyzed using Michaelis-Menten kinetic modeling to determine uptake rates and release concentrations. SI animals show significant increases in stimulated DA release and uptake over GH (Figure 1A and B; n = 8; p < 0.01). Paired pulse experiments examining amplitude of secondary peaks revealed an interaction between stimulation intervals and housing, with increased peaks for SI rats at longer intervals (Figure 1C; n = 6; p < 0.05). Lastly, quinpirole decreased stimulated DA release similarly in both SI and GH rats, suggesting that housing rearing produces no differences in DA autoreceptor activity (Figure 1D; n = 5).

CONCLUSIONS: We have shown that SI rearing increases DA release, uptake, and paired pulse DA release, with no change in autoreceptor activity. Greater DA release may be partly responsible for disrupted DA related behaviors in isolates, including increased drug self-administration, and schizophrenia related behaviors.

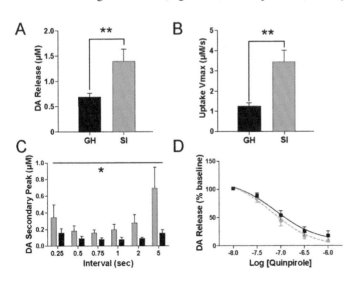

Modulation of Inflammatory Cytokine Release by Hypoxia Mediated Tyrosine Hydroxylase Induction in Mixed Synovial Cells of Patients with Rheumatoid Arthritis and Osteoarthritis

Zsuzsa Jenei-Lanzl[1], Silvia Capellino[1], Frieder Kees[2] and Rainer H. Straub[1]

[1]*University Hospital Regensburg, Regensburg, Bavaria, Germany;* [2]*University of Regensburg, Regensburg, Bavaria, Germany*

BACKGROUND: It is known that the microenvironment of inflamed joints is hypoxic and that hypoxia induces tyrosine hydroxylase (TH) in vivo. Furthermore, in previous studies we have shown that TH-positive, catecholamine-producing cells are present in inflamed synovial tissue (Capellino et al, 2010). In addition, a further study showed that the sympathetic nervous system has a dual role in RA (Haerle et al. 2005).

OBJECTIVES: Therefore, the aim of our study was to investigate whether hypoxia is responsible for TH expression and how hypoxia-induced catecholamines influence the inflammatory response in arthritis.

METHODS AND RESULTS: Synovial tissue was obtained from rheumatoid arthritis (RA) and osteoarthritis (OA) patients. Synovial cells were isolated by enzymatic digestion and cultivated under normoxic or hypoxic conditions. After 24 hours, cells were stained for TH. In order to determine effects caused by TH, cells were incubated with the competitive TH-inhibitor alpha-methyl-p-tyrosine (αMPT). In addition, different concentrations of catecholamine receptor (α- and β-adrenergic, D1/D5, D2,D3,D4 dopamine) or adenosine (A1aR, A2aR) receptor antagonists were applied. Furthermore, cofactors required for optimal TH activity were applied additionally. After 24 hours, supernatants were collected and concentrations of released cytokines (TNF, IL-6, IL-8, IL-10) and catecholamines were determined. Hypoxia increased the number of TH positive cells both in OA and RA compared to cells cultured under normoxia (Figure 1A). Compared to normoxic conditions, hypoxia exhibited inhibitory effects on IL-6, IL-8, and IL-10, whereas TNF was unaffected in OA (Figure 1B). In contrast, hypoxia increased IL-6 and IL-8, but inhibited IL-10 and TNF in RA (Figure 1B). TH blockade by αMPT reversed hypoxia-induced effects on TNF. Moreover, specific catecholamine receptor antagonists (β2-adrenergic, D1/D5 dopamine, A2aR) were able to reverse hypoxia-induced influences on cytokines, especially on TNF, in both OA and RA cells. Additional incubation with specific TH cofactors emphasized the hypoxia-induced effects on TNF (Figure 1C). The very first HPLC measurements showed low catecholamine concentrations.

CONCLUSIONS: This study demonstrates that hypoxia induces TH expression and inhibits TNF in synovial cells, especially in RA patients. In summary, these results suggest that hypoxia influences the inflammatory response in RA synovial cells. The role of catecholamines still has to be defined. In future experiments we will test the effect of local injected TH positive cells in mice with collagen induced arthritis.

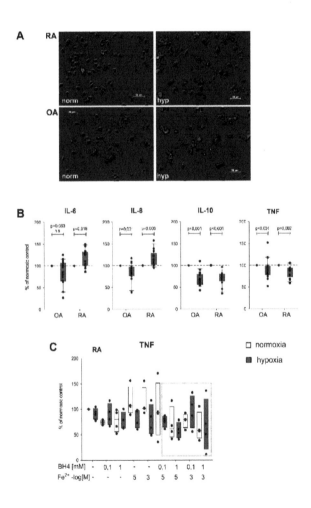

Catecholamine Interactions with Other Transmitters

J

Lee Eiden and David Goldstein

The final Theme of the symposium, like the previous one centered on integrative aspects of catecholaminergic function, in this case catecholamine interactions with other neurotransmitter systems. Two major areas of emphasis were the role of serotonin-catecholamine interactions in central nervous system function and dysfunction, and the coordination of stress responses in brain and periphery by catecholaminergic modulation of peptidergic systems, and vice versa.

Theme J began with a session on modulation of noradrenergic neurotransmission by adenosine, PACAP, and angiotensin, co-chaired by Thomas Westfall (St. Louis University) and Juan Saavedra (NIMH) and convened in honor of Zofia Zukowska, a distinguished researcher in metabolic disorder and stress, and an organizer of the Tenth International Catecholamine Symposium who passed away just months before the meeting in Asilomar in September, 2012. Dr. Zukowska's recent work revealed the direct role of NPY in mediating stress-induced obesity associated with sympathetic overactivity. The contributions of the laboratories of Westfall, Cavadas, Shioda, Saavedra and

Sabban appropriately illustrated the primacy of sympathetic drive mediating the major physiological acute stress response as well as pathophysiological aspects of chronic stress, and therefore the importance of modulation of this primary response by ancillary neuropeptide systems supporting in some cases, and opposing in other cases, sympathetic and adrenal activation. Synergistic, antagonistic, and permissive effects of PACAP, angiotensin, NPY and adenosine can all be expected to be potential avenues for pharmacological modulation of the stress response. The challenge of attenuating stress pathology, while preserving the necessary primary homeostatic stress response, was a clear emergent message of this session.

The session 'Catecholamines and other transmitters in stress' again emphasized the reciprocal relationships between catecholamines and neuropeptides in the stress response, but featuring non-catecholamine transmitters dominant in the primary stress response, in some cases dependent on the stress modality under study. Noradrenergic cells in the locus coeruleus (LC) are situated to respond to the sensory inputs associated with stress, and to modulate both cortical and hypothalamic centers in a stress-dependent way. As described by Van Bockstaele and Valentino, neuropeptides including opioids and CRF modulate LC activity via complex presynaptic mechanisms. In the case of CRF, its modulation of noradrenergic function is directly reciprocal, since the LC is a major regulation of the activity and biosynthetic capacity of the CRF neuron of the paraventricular hypothalamus (PVH)-the final common pathway for activation of the HPA axis leading to elevation of corticosterone/cortisol during stress. Watts and Kahn demonstrated this regulation to be specific to systemic stressors, such as hypoglycemia. Mustafa marshaled recent evidence for a reciprocal role of PACAP at the level of the PVN to that of norepinephrine. PACAP-deficient mice exhibit attenuation of HPA axis activation by *psychological* stress, with no apparent role in mediating the effects of *systemic* stress, while in the periphery, PACAP is the major splanchnicoadrenomedullary transmitter responsible for catecholamine release from chromaffin cells in response to *either* systemic or psychological stress (see also C. Smith, Theme B). Gary Thomas's elegant molecular biological dissection of modulation of memory consolidation and retrieval in genetically altered mice during stress reveals that noradrenergic mechanisms are essentially redundant with dopamine D5-receptor-dependent ones, revealing a novel point of interaction between the nominally 'rewarding' and nominally 'arousing' dopamine and norepinephrine systems during processing of aversive stimulation.

The final session of the meeting and of Theme J was chaired by Julie Hensler, who with P. De Deurwaerdere, A. Bortolozzi, and L. Daws addressed tough and clinically fraught questions about the relative roles of 5-HT and NE in treatment of depression; the functional selectivity in vivo of SSRIs on serotonin function during treatment for depression, and the unresolved roles of anxiety and anhedonia and their interaction in various forms and symptoms of clinical depression.

Catecholamine Release Modulation by Adenosine through A2AReceptors in Mouse Chromaffin Cell Culture

Magda M. Santana[1,3], Joana Rosmaninho-Salgado[1], Manuella Kaster[1], Vera Cortez[1], Rodrigo A. Cunha[1,2] and Cláudia Cavadas[1,3]

[1]*Center for Neuroscience and Cell Biology, University of Coimbra, Coimbra, Portugal;* [2]*Institute of Biochemistry, Faculty of Medicine, University of Coimbra, Coimbra, Portugal;* [3]*Faculty of Pharmacy, University of Coimbra, Coimbra, Portugal*

BACKGROUND: Sympathetic nervous system and adrenal medulla secrete high amounts of catecholamines (norepinephrine, NE; and epinephrine, EP) that are one of the primary mediators for many ofthe physiological consequences of the response to stress. Chromaffin cells of the adrenal medulla are the main peripheral component involved in catecholamine release, especially EP. The adenosinergic system is a neuromodulatory system able to control the activity of other neurotransmitters and hormonal systems strongly related to stress responses. The concentration of adenosine increases during stress and when released into, or formed in, the extracellular space, adenosine binds to four types of G protein-coupled receptor, A1, A2A, A2B, and/or A3receptors (Fredholm BB., 2006). The role of adrenal adenosine receptors on catecholamine release is not completely known.

OBJECTIVES: The aim of our work was to study the role A1and A2Aadenosine receptors on catecholamine release from adrenal chromaffin cells using primary cell cultures of mouse adrenal chromaffin cells.

METHODS AND RESULTS: Cells were incubated for 5 min with Krebs containing the drugs tested. The levels of norepinephrine (NE) and epinephrine (EP) released in the medium were determined by HPLC with electrochemical detection. By immunohistochemistry adenosine A2Areceptors were found in the medulla of mouse adrenal gland. On contrary, adenosine A1 receptors were not found in the adrenal medulla, but immunoreactivity for these receptors was observed in the reticularis adrenal cortical layer. The A2Aadenosine receptor activation of mouse chromaffin cell by incubating chromaffin cells with CGS21680 (30 nM, A2Areceptor agonist) stimulated NE and EP release. This effect was blocked by the A2Areceptor antagonist (SCH58261, 50 nM), chelerytrine (6 μM, a PKC inhibitor), H-89 (2 μM, a PKA inhibitor) or AMPT (1 μM, a tyrosine hydroxylase, TH, inhibitor), suggesting that A2Areceptor stimulated catecholamine release by a mechanism dependent on PKA, PKC and TH activation. The A1receptor agonist (CPA, 100 nM) did not change catecholamine release from mouse cromaffin cells.

CONCLUSIONS: Our results show that A2Aadenosine receptors modulate catecholamine release, suggesting a potential of A2Areceptors antagonists to manage modifications induced by stress.

Support by FCT and FEDER: PTDC/SAU-NEU/108110/2008, SFRH/BPD/31547/2006, SFRH/BD/44664/2008.

Morphological and Functional Significance of Coexistence of PACAP with Catecholamine in Nerve Terminals Innervating Hypothalamic Neuroendocrine Cells

Seiji Shioda[1], Tomoya Nakamachi[1], Keisuke Kiriyama[1,2] and Toshihiko Yada[3]

[1]*Showa University Tokyo, Tokyo, Japan;* [2]*For Days Co., Ltd Tokyo, Tokyo, Japan;* [3]*Jichi University Tochigi, Tochigi, Japan*

BACKGROUND: Pituitary adenylate cyclase-activating polypeptide (PACAP) is a pleiotropic neuropeptide and its receptors are widely distributed in brain including the hypothalamus. Many PACAP-containing nerve terminals are shown to innervate and activate arginine-vasopressin (AVP)-containing neurons in the rat hypothalamus. Noradrenalin (NA) activates AVP-containing neurons via α1-receptor, which is partly linked to Ca2+ signaling partly via a cAMP-PKA-regulated pathway. We have already shown that PACAP and catecholamine are coexisted in the same nerve terminals in the hypothalamus, so they are suggested to originate from the caudal ventrolateral medulla (CVLM). As the pathway arising from PACAP/NA-containing neurons in the CVLM is involved in the regulation of AVP neurons, we studied the physiological significance of the coexistence of PACAP with NA in the same nerve terminals.

OBJECTIVES: We first developed a double immunostaining method by use of antisera against PACAP and catecholamine-synthesizing enzyme such as tyrosine-hydroxylase (TH) and dopamine β-hydroxylase (DBH) to identify the coexistence of PACAP and NA in the same nerve terminals. Furthermore, we isolated neurosecretory cells from the rat hypothalamus to study the functional significance of its coexistence by use of a physiological method.

METHODS AND RESULTS: Adult male SD rats were used in this study. The animals were sacrificed and the tissue sections were immunostained with PACAP or TH and DBH antisera and ABC method. A pre-embedding double immunostaining method was used at light and electron microscopic level. Some ultrathin sections were immunostained with post- embedding immunocolloidal gold staining with antibody against neurophysin II (NPII) which is a carrier protein of AVP. The cytosolic Ca was measured by the fluorescence image from the fura-2-loaded cells which were isolated from rat hypothalamus. The neurosecretory cells isolated from the rat hypothalamus were loaded with the fluorescent dye quinacrine and fluorescence was recorded and emission was recorded. PACAP immunoreactivity was detected in the granular vesicles in catecholaminergic nerve terminals that made synaptic contact with AVP neurons. Both PACAP and NA induced large increases in the cytosolic Ca2+ concentration in isolated AVP cells. Lower concentrations of

PACAP and NA had little effects in separately, however, when both PACAP and NA were combined, they evoked large increase in cytosolic Ca2 + in AVP neurons. In AVP cells that were prelabeled with quinacrine, PACAP and NA acted synergistically to induce a loss of quinacrine fluorescence, suggesting secretion of neurosecretory granules in AVP neurons.

CONCLUSIONS: These results may suggest that PACAP and NA, co-released from the same nerve terminals, act in synergy to evoke calcium signaling and secretion in AVP neurons. The synergism is mediated by the interaction between cAMP-PKA pathway and with an as yet unidentified factor "X" linked to L-type Ca2 + channels. The synergism between PACAP and NA may contribute to the regulation of AVP secretion under physiological conditions.

Interactions of Neuropeptide Y, Catecholamines and Angiotensin at the Vascular Neuroeffector Junction

Thomas C. Westfall, Heather Macarthur, Mirnela Byku and Jessica Murray

Department of Pharmacological and Physiological Science, Saint Louis University School of Medicine, St. Louis, MO 63104, USA

The sympathetic nervous system and renin-angiotensin system are both thought to contribute to the development and maintenance of hypertension in experimental models such as the Spontaneously Hypertensive Rat (SHR). We demonstrated that periarterial nerve stimulation (NS) increased the perfusion pressure (PP), NE overflow and neuropeptide Y (NPY) overflow from the perfused mesenteric arterial beds of SHR and WKY at 4−6, 10−12 and 18−20 wk of age. The effects were greater in vessels obtained from SHR of all 3 ages compared with WKY. Angiotensin II (Ang II) produces an enhancement of the NS-induced increase in the release of NPY and NE and increase in PP in the mesenteric bed obtained from SHR and the effect was greater in beds obtained from SHR compared to age matched controls. In contrast Ang (1−7) decreased NS-induced release of NE and NPY as well as the increase in PP. This effect was greater in arteries obtained from SHR than WKY. A MAS receptor antagonist attenuated the Ang (1−7) induced decrease in both NE and NPY release while an AT2 receptor antagonist only affected the Ang (1−7) induced decrease in NE without affecting the decrease in NPY overflow. Both an ACE inhibitor and an AT1 receptor blocker decreased the NS-stimulation induced overflow of NE and NPY. Plasma Ang II levels are greater in the 10−12 wk SHR than WKY while plasma Ang (1−7) levels are lower in the SHR than WKY. Taken together the opposing effects of Ang (1−7) and Ang II suggest that there is a normal balance between these two hormones in modulating sympathetic neurotransmission and an imbalance may contribute to the development and maintenance of hypertension.

Overview: Neuropeptide-Catecholamine Interactions in Stress

Lee Eiden

NIMH-IRP, USA

The session "Neuropeptide-catecholamine interactions in stress" provides a very current snapshot of the central and peripheral circuits mediating stress responses in rodents, and its relevance to stress-related human disorders including PTSD, depression, and anxiety, as well as allostatic effects of chronic stress on cardiovascular and cognitive functioning. Beginning in the periphery, sensory and systemic inputs to brainstem nuclei converge on the paraventricular nucleus (PVN) to activate the hypothalamo-pituitary-adrenal (HPA) axis. At the same time, modulation conveyed via the locus coeruleus of more rostral brain areas, including hippocampus, frontal cortex, and extended amygdala, as well as past experiences embedded as response biases within those areas, modulate HPA axis activity by both direct and indirect inputs to the PVN. Finally, pituitary and spinal pre-sympathetic outputs to the adrenal cortex and medulla, respectively regulate peripheral hormonal secretion in response to stress, either systemic or psychogenic (a.k.a. 'physiological' or 'psychological'). It is striking that catecholamine regulation of peptide hormone secretion (e.g. noradrenergic regulation of CRH release) and peptide hormone regulation of catecholamine secretion (e.g. enkephalin and CRH control of locus coeruleus activity; PACAP regulation of adrenomedullary epinephrine secretion) are intercalated within central and peripheral stress circuitry. Multiple modes of signaling via parallel receptor activation following either neuropeptide or catecholamine release in the brain provide an additional level of conditional signaling complexity for highly contingent activation of these circuits.

Neuropeptide Regulation of the Locus Coeruleus (LC) in the Mediation of Stress Responses

Elisabeth J. Van Bockstaele and Rita J. Valentino

Department of Neuroscience, Farber Institute for Neurosciences, Thomas Jefferson University, Philadelphia PA 19107; Department of Pediatrics, Division of Stress Neurobiology, Children's Hospital of Philadelphia, Philadelphia, PA 19103, USA

The LC is a compact, homogeneous norepinephrine (NE)-containing nucleus that innervates the entire neuraxis through a divergent efferent system. It is the sole source of NE in many forebrain regions that have been implicated in cognition (e.g., cortex and hippocampus). LC neurons are spontaneously active and their rate of discharge is positively correlated to behavioral and electroencephalographic indices of arousal. The LC is finely tuned by co-regulation between the endogenous opioids, enkephalin (ENK) and dynorphin (DYN), and corticotropin releasing factor (CRF). Work from our laboratories has revealed that during stress (physiological or psychological), CRF is released (likely from the central nucleus of the amygdala CNA) to shift the activity of LC neurons to a high tonic state that would promote scanning of the environment and behavioral flexibility. At the same time, endogenous opioids acting at m-OR in the LC (via ENK) exert an opposing inhibitory effect that may serve to restrain the excitatory actions of CRF and help to bring neuronal activity back to baseline. The CRF and ENK that regulate the LC derive from distinct sources (CNA and medulla, respectively) but their axon terminals converge onto common LC neurons which can respond to both peptides because they co-express m-OR and CRF receptors. Both ENK and CRF axon terminals co-localize glutamate, which mediates the short-lived LC activation by sensory stimuli. Most recently, we demonstrated that the dynorphin-k-OR receptor system exerts another layer of regulation on the LC system by presynaptic inhibition of excitatory LC afferents. Dynorphin was found to be co-localized with CRF and k-OR was localized to CRF, glutamate and DYN-containing axon terminals. Electrophysiological studies demonstrated selective presynaptic inhibition of glutamatergic and CRF afferent input by selective k-OR agonists. By allowing LC neurons to fire spontaneously, but attenuating information from excitatory afferents, the dynorphin-k-OR system takes the LC off-line, preventing adaptive responses to sensory stimuli or stressors. This may serve to protect the LC from over-activation. However, it might also be predicted to promote passive behavior as is characteristic of depression. Given how LC activity is finely tuned by the integration of CRF and endogenous opioid inputs, upsetting the CRF:opioid balance in the LC could influence the stress-sensitivity of this system or its sensitivity to opiates. Evidence will be presented regarding the importance of maintaining appropriate interactions between ENK-m-OR, DYN-k-OR and CRF in the LC and the pathological consequences that can develop from a dysregulation of these systems.

Breaking Down Monoamine "Silos": We're all in this Together

Julie Hensler

Department of Pharmacology, University of Texas Health Science Center, San Antonio, TX, USA

An "information silo" is an isolated system, unable to exchange information with other, related systems. The "silo effect" refers to a lack of communication and common goals between departments within an organization, or in our case, disciplines or fields of study. The monoaminergic systems, serotonin (5-HT), norepinephrine (NE) and dopamine (DA), are often viewed in isolation, i.e. as separate and distinct entities. However, neuroanatomical, neurochemical and pharmacological evidence is to the contrary. The organization of these transmitter systems within the central nervous system and in the periphery supports the idea of a wonderfully intricate and complex scheme of reciprocity. In this session we invite "systems thinking", the process of understanding how monoamines influence one another within the whole system, whether it is the central or enteric nervous systems. We have organized a session to bring together cross-functional thought leaders in areas related to schizophrenia, Parkinson's disease, major depression and gastrointestinal function. Transporters for 5-HT, NE and DA are the key players in terminating transmission of these amines through high-affinity uptake. Promiscuity among these transporters in the central nervous system (the uptake of multiple amines in addition to their "native" transmitter) may influence the therapeutic efficacy uptake blockers, and the DA precursor L-DOPA. The less than hoped for antidepressant efficacy of currently prescribed uptake inhibitors may be a result of this promiscuity, and the action of "non-traditional" transporters such as organic cation transporters and the plasma membrane monoamine transporter. The mechanism of action of L-DOPA in the management of Parkinsons' disease, and both motor and non-motor complications of L- DOPA therapy, may be determined by interactions between the 5-HT, DA and NE systems. A greater appreciation of interactions between monoaminergic systems in cortical areas may provide opportunities for improve therapies in the treatment of schizophrenia. In the periphery, interactions between 5-HT and DA neurons of the enteric nervous system affect gastrointestinal function. A greater understanding of the intricate interactions between 5-HT, and DA/NE can be incorporated into better strategies for the treatment of gastrointestinal dysfunction, and in the development of more effective therapies for disorders such as major depression, Parkinson's disease, and schizophrenia.

Orexin/hypocretin in the Ventral Tegmental Area is Necessary for Morphine-Induced Synaptic Plasticity of Dopamine Neurons

Corey Baimel[1] and Stephanie L. Borgland[1]

[1]*University of British Columbia, Vancouver, British Columbia, Canada*

BACKGROUND: Dopamine neurons in the ventral tegmental area (VTA) are a key target of addictive drugs and neuroplasticity in this region may underlie some of the core features of addiction. All drugs of abuse induce an LTP-like potentiation of excitatory inputs to VTA dopamine neurons. Orexin/hypocretin (orx/hcrt) is a lateral hypothalamic neuropeptide released into the VTA that exerts modulatory effects on a variety of behaviors produced by drugs of abuse. Acute application of orx/hcrt potentiates excitatory synaptic transmission in the VTA, and inhibition of orx/hcrt signaling blocks both cocaine-induced plasticity and behavioral sensitization. However, the role of orx/hcrt on the plasticity induced by other classes of abused drugs is unknown.

OBJECTIVES: Here we aimed to test if orexin/hypocretin action was necessary for morphine-induced synaptic plasticity of VTA dopamine neurons.

METHODS AND RESULTS: Morphine-induced plasticity of VTA dopamine neurons was assessed 24 hrs following a single *in vivo* injection of morphine (10 mg/kg) or saline, using whole-cell patch clamp electrophysiology. To test for orx/hcrt involvement in morphine-induced plasticity, the orx/hcrt type 1 receptor antagonist SB 334867 (10 mg/kg) was delivered both systemically and directly into the VTA prior to morphine treatment. Morphine potentiated glutamatergic synapses by a pre-synaptic increase in glutamate release and by a post-synaptic change in AMPAR number or function, likely including a switch in subunit composition. Systemic and intra-VTA administration of SB 334867 blocked a morphine-induced increase in the AMPAR/NMDAR ratio, morphine-induced increases in AMPAR mEPSC frequency and amplitude, as well as morphine-induced AMPAR redistribution measured by a change in rectification.

CONCLUSIONS: These results support a role for orx/hcrt signaling in both pre-and post-synaptic potentiation of glutamatergic transmission in the VTA by morphine. Because orx/hcrt signaling is required for plasticity induced by both morphine and cocaine, orx/hcrt may function as a gatekeeper for drug-induced plasticity of dopamine neurons.

The Adrenomedullary Angiotensin II Type 2 Receptor: Stress-Triggered Regulation and Role in Catecholamine Biosynthesis

R. Nostramo[1], A. Tillinger[1], J.M. Saavedra[2], L. Serova[1], R. Kvetnansky[3] and E.L. Sabban[1]

[1]*Department of Biochemistry and Molecular Biology, New York Medical College, Valhalla, New York 10595, USA;*
[2]*Section of Pharmacology, DIRP, National Institute of Mental Health, NIH, Bethesda, MD 20892, USA; [3]Institute of Experimental Endocrinology, Slovak Academy of Sciences, Bratislava, Slovakia*

BACKGROUND: While the renin-angiotensin system (RAS) is important for adrenomedullary responses to stress, the involvement of specific angiotensin II receptor subtypes is unclear.

OBJECTIVES AND METHODS: To better understand the role of the RAS in the stress response, we examined the kinetics of angiotensin II type 1A receptor ($AT_{1A}R$) and type 2 receptor (AT_2R) gene expression changes in the adrenal medulla of male intact or hypophysectomized Sprague Dawley rats or corticotropin-releasing hormone knockout mice in response to single and repeated immobilization stress (IMO) and characterized the role of these receptors on catecholamine biosynthesis in PC12 cells.

RESULTS: AT_2R mRNA levels decreased rapidly by 90% after a single 2 h IMO. Repeated IMO also decreased AT_2R mRNA levels, but the decline was more transient. $AT_{1A}R$ mRNA levels were unaltered with either single or repeated IMO, although binding was increased with repeated IMO. These effects of stress on angiotensin II receptor expression may alter catecholamine biosynthesis, as tyrosine hydroxylase (TH) and dopamine beta-hydroxylase (DBH) mRNA levels in PC12 cells are decreased with angiotensin II treatment in the presence of ZD7155 (AT_1R antagonist), or with CGP42112 (AT_2R agonist). Involvement of stress-triggered hypothalamic-pituitary-adrenocortical or sympatho-adrenal axis activation in AT_2R downregulation was examined. Dexamethasone triggered a transcriptionally-mediated decrease in AT_2R mRNA levels in PC12 cells. However, glucocorticoids are not required for the immediate stress-triggered decrease in AT_2R gene expression, demonstrated in corticotropin-releasing hormone knockout mice and hypophysectomized rats. They can regulate its basal gene expression. cAMP and pituitary adenylate cyclase-activating polypeptide (PACAP) also reduced AT_2R gene expression.

CONCLUSIONS: Changes in $AT_{1A}R$ and AT_2R expression in the adrenal medulla likely contribute to the regulation of adrenomedullary function with stress. Downregulation of AT_2R gene expression does not require the stress-triggered rise in glucocorticoids, but may be mediated by the release of PACAP and subsequent elevation in cAMP.

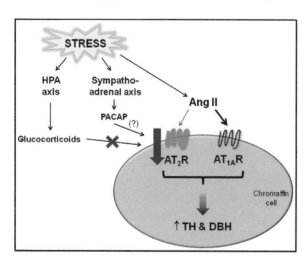

Angiotensin AT$_1$ Receptor Blockade Prevents Excessive Brain and Peripheral Sympathetic Stimulation During Stress

Juan M. Saavedra[1]

[1]*NIMH-IRP, Bethesda, Maryland, United States*

Angiotensin II, by stimulation of its physiological AT$_1$ receptors, contributes to control the basal as well as the stress-induced peripheral and central sympathetic activity. Control mechanisms include the regulation of norepinephrine release from sympathetic nerves, catecholamine production and release from the adrenal medulla, and the central sympathetic activity controlled by the locus coeruleus. Increased central and peripheral AT$_1$ receptor expression and activity is a regular occurrence in many stress models, including isolation, restraint, cold and inflammatory stress. Effective brain and peripheral AT$_1$ receptor blockade is achieved by systemic administration of orally active sartans (Angiotensin II AT$_1$ receptor blockers, ARBs). Sartan administration prevents and decreases excessive central and adrenomedullary catecholamine production and release in many stress models, reduces pathological stress responses and ameliorates stress-induced disorders such as gastric ulceration. Excessive brain AT$_1$ receptor activity not only associates with abnormal stress responses but also with hypertension, brain ischemia, blood brain barrier breakdown and inflammation. Thus, pathological AT$_1$ receptor activity participates in the development of risk factors leading to neuronal injury, the incidence and progression of neurodegerative, mood and traumatic brain disorders and cognitive decline. Consequently, studies in animal models demonstrated that AT$_1$ receptor blockade decreases anxiety and depression, protects cerebral blood flow during stroke, and decreases brain inflammation, amyloid beta neurotoxicity and the neurological deficits following traumatic brain injury. Controlled clinical studies indicate that sartans protect cognition after stroke and during aging. Cohort analyses reveal that these compounds significantly reduce the incidence and progression of Alzheimer's disease. Sartans are commonly used for the therapy of hypertension, diabetes and stroke, but have not been studied in the context of disorders when uncontrolled stress contributes to their initiation and progress. Sartans are well-tolerated pleiotropic neuroprotective compounds with additional beneficial cardiovascular and metabolic profiles. Their use offers a novel therapeutic approach of immediate translational value.

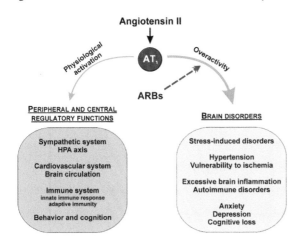

Catecholamine Release Modulation by Adenosine Through A_{2a} Receptors in Mouse Chromaffin Cells in Culture

Magda M. Santana[1,3], Joana Rosmaninho-Salgado[1], Manuella Kaster[1], Vera Cortez[1], Rodrigo A. Cunha[1,2] and Cláudia Cavadas[1,3]

[1]*CNC- Center for Neuroscience and Cell Biology, University of Coimbra, Coimbra, Portugal;* [2]*Institute of Biochemistry, Faculty of Medicine, University of Coimbra, Portugal;* [3]*Faculty of Pharmacy, University of Coimbra, Coimbra, Portugal.*

BACKGROUD: The adrenal medulla secretes high amounts of catecholamine (norepinephrine, NE; and epinephrine, EP) that are one of the primary mediators of the response to stress. Chromaffin cells of the adrenal medulla are the main peripheral component involved in

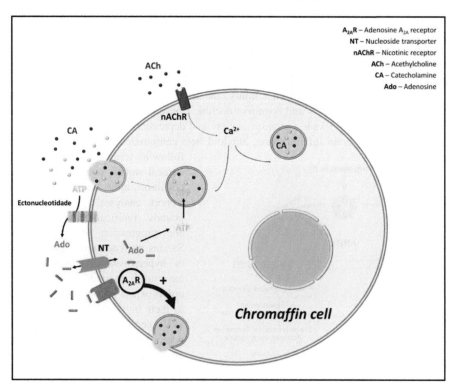

catecholamine release, especially EP. The adenosinergic system is a neuromodulatory system able to control the activity of other neurotransmitters and hormones related to stress, through the activation of four types of G protein-coupled receptor (A_1, A_{2A}, A_{2B}, and/or A_3 receptors). The adenosine receptors can be found within adrenal gland [1], however, their exact localization in adrenal gland and their role on catecholamine release are not completely understood.

OBJECTIVES: To study the localization of A_1 and A_{2A} adenosine receptors in the mouse andrenal gland and their role on catecholamine release from mouse adrenal chromaffin cells.

METHODS AND RESULTS: Primary cell cultures from mouse chromaffin cells [2] were incubated for 5 min with Krebs buffer containing the drugs. The levels of NE and EP released from the cells were determined by HPLC-ED. By immunohistochemistry, the mouse adrenal medulla showed adenosine A_{2A} receptors-immunoreactivity (-ir). On the contrary, the adenosine A_1 receptor-ir was not observed in the adrenal medulla, but only in the inner adrenal cortical layer.

The A_{2A} adenosine receptor activation of mouse chromaffin cell by incubating chromaffin cells with the A_{2A} receptor agonist (CGS21680, 30 nM) stimulated NE and EP release. This effect was blocked by the A_{2A} receptor antagonist (SCH58261, 50 nM), the PKC inhibitor (chelerytrine, 6 μM), the PKA inhibitor (H-89; 2 μM, a). These results suggest that A_{2A} receptors stimulate catecholamine release from mouse chromaffin cells by a mechanism dependent on PKA and/or PKC activation The A_1 receptor agonist (CPA, 100 nM) did not change catecholamine release from mouse cromaffin cells.

CONCLUSIONS: Our results show that A_{2A} adenosine receptors modulate catecholamine release from adrenal chromaffin cells, suggesting the adrenal A_{2A} receptors as targets to manage modifications induced by stress.

Supported by FCT and FEDER: PTDC/SAU-NEU/108110/2008, SFRH/BPD/31547/2006, SFRH/BD/44664/2008.

References

1. Chen *et al*, Int J Biochem Cell Biol 40: 2815−25
2. Cavadas *et al*, PNAS 103:10497−10502, 2006

Pituitary Adenylate Cyclase-Activating Polypeptide (PACAP), A Master Regulator in Central and Peripheral Stress Responses

Tomris Mustafa[1], Nikolas Stroth[1], Michael Lehmann[1], Shyue-An Chan[2], Corey B. Smith[2] and Lee E. Eiden[1]

[1]*Sections on Molecular Neuroscience and Functional Neuroanatomy, NIMH-IRP, Bethesda, MD;* [2]*Department of Biophysics, Case Western Reserve University, Cleveland, OH, United States.*

BACKGROUND: Mammals respond to systemic and psychogenic stress through increased splanchnic nerve-driven adrenal catecholamine secretion and hypothalamic corticotropin-releasing hormone (CRH)- mediated corticosterone (CORT) secretion from the adrenal cortex. Work from our laboratory and others implicates the neuropeptide PACAP in both responses.

OBJECTIVES: We have hypothesized that PACAP is required for the transduction of the stress response at two separate anatomical loci: the splanchnic-adrenal synapse and the hypothalamo-pituitary-adrenal (HPA) axis. We examined the impact of PACAP deficiency on control of adrenal catecholamine secretion and biosynthesis (stimulus-secretion-synthesis coupling), and HPA responses, to chronic psychogenic and systemic stressors.

METHODS AND RESULTS: Systemic (insulin-induced hypoglycemia) or psychological/ psychogenic (restraint) stress caused an~2-fold induction in both tyrosine hydroxylase (TH), and phenylethanolamine N-methyltransferase (PNMT) mRNA in the adrenal that was blunted or absent in PACAP-deficient mice. Catecholamine secretion following high-frequency stimulation (15 Hz) of the splanchnic nerve, in ex vivo adrenal slice preparations, was essentially abolished in slices from PACAP-deficient mice. CORT elevation after 6 hours of restraint stress was grossly attenuated in PACAP-deficient mice, while CORT elevation after administration of lipopolysaccharide (LPS) was not. PACAP-dependent CORT elevation in acute restraint stress was more pronounced at longer durations of restraint ($>60\%$ at 6 hrs, compared to 40% at 1 hr), suggesting that PACAP's role in central stress transduction might be even greater in chronic, compared to acute, stress. This hypothesis proved correct: after two weeks of social defeat, there was no significant increase in CORT elevation measured 30 minutes post-defeat in PACAP-deficient mice. Furthermore, PACAP deficiency offered protection against the development of anxiety- and depressive-like behaviors elicited by two weeks of chronic social defeat.

CONCLUSIONS: This work identifies PACAP as the autonomic neurotransmitter mediating splanchnic-adrenal catecholamine secretion independently of stressor modality and as a central neurotransmitter mediating CORT secretion in response to psychogenic but not systemic stress. PACAP antagonists may have therapeutic value for the treatment of hypercortisolemia associated with chronic stress and depression.

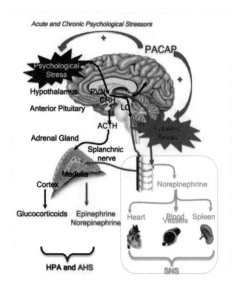

Novel Catecholamine Mechanisms in Memory Consolidation and Retrieval During Stress

Steven A. Thomas, Ming Ouyang, Keith B. Schutsky, Matthew B. Young, Christina B. Castelino, Lei Zhang and Melissa M. Lestini

Pharmacology, University of Pennsylvania, Philadelphia, PA USA

BACKGROUND: A widely held view is that the stress-responsive adrenergic system is critical to the consolidation of memory for emotional experiences. For example, stimulating β-adrenergic receptors shortly after an experience can enhance long-term memory for that experience. However, genetic evidence indicating that adrenergic signaling is required for emotional memory consolidation is lacking. Mice in which the adrenergic ligands norepinephrine and epinephrine (NE/E) are absent due to targeted disruption of the dopamine β-hydroxylase gene (*Dbh*) exhibit intact consolidation of fear memory. In contrast, they exhibit a deficit in the retrieval of hippocampus-dependent memory that is mediated by β_1-adrenergic receptors. However, it has been proposed that stress and glucocorticoids impair memory retrieval by enhancing NE/E signaling via β_1 receptors.

OBJECTIVES: The current studies were undertaken to resolve these discrepancies.

METHODS AND RESULTS: Mice and rats were trained using single-trial classical or instrumental fear. Genetic and pharmacologic approaches were used, either shortly after training (consolidation), or shortly before testing (retrieval). Corticosterone impaired contextual and instrumental fear memory retrieval in the absence of either NE/E (*Dbh*$^{-/-}$ mice) or the β_1 receptor (β_1 KO mice), and in the presence of a β_1 antagonist (WT mice). In contrast, corticosterone did not impair retrieval in WT mice treated with a β_2 antagonist or in β_2 KO mice. Further, β_2 agonists impaired retrieval in WT mice. The impairing effect of restraint stress on retrieval was blocked by a β_2 antagonist or in β_2 KO mice. The impairing effects of stress, corticosterone or a β_2 agonist were also blocked by pretreatment of the dorsal hippocampus with pertussis toxin, suggesting that β_2 receptors couple to G_i. In support of this, selectively stimulating β_2 receptors in slices from the dorsal hippocampus caused cAMP levels to fall, and receptor co-stimulation prevented the rise in cAMP caused by selectively stimulating β_1 receptors.

In contrast, cued and instrumental fear memory consolidation were normal in *Dbh*$^{-/-}$ mice. However, dopamine $D_{1,5}$ receptor antagonists impaired consolidation of cued fear in *Dbh*$^{-/-}$ mice, β_2 KO mice, and WT mice if they were also treated with a β_2 antagonist. Further, a β_2 antagonist impaired consolidation in D_5 but not D_1 KO mice. Infusion of a phospholipase C (PLC) agonist or antagonist into the basolateral amygdala (BLA) enhanced or impaired consolidation, respectively. Measurement of IP$_3$ in the BLA indicated that β_2 and D_5 receptors both signal through PLC to promote consolidation. BLA pretreatment with pertussis toxin indicated that β_2 but not D_5 receptors couple to G_i to activate PLC.

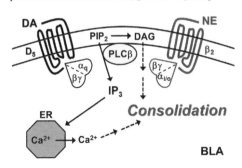

CONCLUSIONS: Dopamine through D_5 receptors signals in a redundant manner with NE/E through β_2 receptors to promote fear memory consolidation. This is mediated in the BLA via mutual receptor activation of PLC rather than adenylyl cyclase. Further, stimulation of β_2 receptors actually decreases levels of cAMP in the hippocampus, an effect that underlies the impairment of memory retrieval mediated by stress and glucocorticoids. The consolidation and the retrieval effects of β_2 receptors are mediated by their coupling to G_i rather than to G_s. The findings demonstrate the need to revise the notion that adrenergic signaling is uniquely required for aversive memory consolidation through activation of β receptors and cAMP signaling.

Catecholaminergic Regulation of Peptide Synthesis and Release Programs in Paraventricular CRH Neurons

Alan G. Watts

*Dept. of Biological Sciences, USC Dornsife College of Letters, Arts and Sciences, University of Southern California,
Los Angeles CA 90089*

BACKGROUND: Hindbrain catecholaminergic neurons are recognized as key controllers of neuroendocrine function. They provide one of the densest inputs to the paraventricular nucleus (PVH), perhaps the most important integrative cell group in the hypothalamus. CRH neuroendocrine neurons in the PVH are the head of the adrenocortical stress response. They control ACTH release from the anterior pituitary gland, a process that is strongly influenced catecholaminergic inputs through their interactions with a local GABA-glutamatergic network. CRH neurons release ACTH secretagogs from median eminence terminals in a stimulus-dependent manner, a process that relies entirely on regulated changes in membrane potentials and ensuing action potentials. But because they release relatively large quantities of peptide into the vasculature, their ability to coordinate peptide synthesis and release in a stimulus-dependent manner is a critical, but rather under-appreciated aspect of their neurobiology.

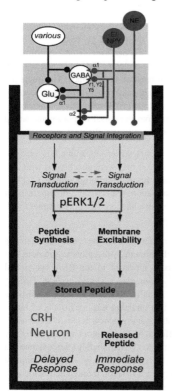

METHODS AND RESULTS: We have investigated how catecholaminergic inputs to control synthesis and release programs in CRH neurons using glycemic stimuli as specific activators. Results from *in vivo* and *ex vivo* studies has focused attention on a MAP kinase cascade involving phospho-ERK1/2 in CRH neurons that is stimulated by catecholamines in an α1 adrenoreceptor-dependent manner, and is strategically positioned to couple peptide synthesis and release in response to afferent signals. 1) pERK1/2 responses closely track ACTH release and *Crh* transcription following intravenous insulin-induced hypoglycemia (IIH) or 2-deoxy-D-glucose (2DG); 2) a catecholaminergic afferent dependency of ACTH, phospho-ERK1/2, and *Crh* transcriptional responses to IIH and 2DG; 3) pERK1/2 retains its sensitivity to other challenges in catecholaminergic afferent-lesioned animals; and 4), ERK1/2 phosphorylation is required for norepinephrine to drive mechanisms that control peptide release (neuronal firing rate) and transcription (phospho-CREB accumulation).

CONCLUSIONS: These results establish clear structural and functional relationships linking catecholaminergic neurons with intracellular mechanisms in CRH neuroendocrine neurons that initiate adrenocortical responses to glycemic challenges.

Supported by R01 NS029728 (NIH) & JDRF 2008-710-03.

Control of Dopamine Neurotransmission in the Prefrontal Cortex: Focus on Serotonin and Norepinephrine Systems

Analía Bortolozzi

IIBB-CSIC-IDIBPAS, Cibersam, Barcelona, Spain

BACKGROUND: Monoamines modulate prefrontal cortex (PFC) function, and their dysregulation plays an important role in a number of psychiatric and neurological disorders. We have focused on serotonin (5-HT) and norepinephrine (NE) pathways to selectively enhance mesocortical dopamine (DA) neurotransmission, since a cortical DA hypofunction may underlie negative symptoms and cognitive deficits in schizophrenia. On the other hand, atypical antipsychotic drugs (APDs) used in the treatment of schizophrenia, target multiple monoamine receptors. However, the mechanism of action of these drugs is still poorly understood, in particular regarding the neurotransmitter interactions involved.

OBJECTIVES: To review the mechanism of action of APDs with a particular emphasis on their action in the prefrontal cortex.

METHODS AND RESULTS: APDs may restore cortical DA activity by preferentially targeting $5\text{-HT}_{2A/2C}R$ and $5\text{-HT}_{1A}R$ receptors in the rodent medial PFC. Histological studies have shown that almost 50% and 20% of the pyramidal- and GABAergic neurons, respectively express both $5\text{-HT}_{2A}R$ and/or $5\text{-HT}_{1A}R$ transcripts in deep layer of medial PFC. Moreover, track-tracing techniques combined with *in situ* hybridization have revealed that a large proportion of pyramidal neurons projecting to the ventral tegmental area (VTA) express $5\text{-HT}_{2A}R$ mRNA, which provides an anatomical substrate for 5-HT-DA relationships. Non-selective $5\text{-HT}_{2A/2C}R$ agonist DOI (40 pmol) infusion in medial PFC increased burst firing of VTA DA cells and mesocortical DA output. Likewise, the activity of VTA DA neurons and local DA release were increased by selective $5\text{-HT}_{1A}R$ agonists in medial PFC. Further, atypical antipsychotic drugs (clozapine, olanzapine, aripiprazole and ziprasidone, $100-300\ \mu M$)-but not haloperidol- increased PFC DA release by $5\text{-HT}_{1A}R$-mediated mechanism. Hence, cortical $5\text{-HT}_{1A}R$ are necessary to modulate the mesocortical DA neurotransmission, an absent effect in $5\text{-HT}_{1A}R$ knockout mice.On the other hand, extracellular DA in medial PFC—but not nucleus accumbens— arises mainly from NE axons and is sensitive to noradrenergic drugs, which offers additional ways to selectively modulate cortical DA activity. Hence NET inhibitor reboxetine (1-10-30 μM) enhanced DA release only in mPFC, an effect augmented by alpha2-adrenocepotor blockade with RX-821002 and the antidepressant drug mirtazapine.

CONCLUSIONS: Overall, these observations support the therapeutic usefulness of 5-HT/NE drugs or drug combinations devised to selectively augment DA transmission in the PFC, circumventing the existing limitations in the treatment of DA-related diseases.

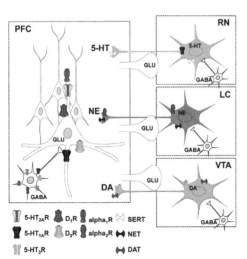

Unfaithful Neurotransmitter Transporters: Focus on Biogenic Amine Reuptake and Implications for Antidepressant Efficacy

Lynette C. Daws[1,3] and Wouter Koek[2,3]

[1]*Departments of Physiology;* [2]*Psychiatry;* [3]*Pharmacology,University of Texas Health Science Center at San Antonio, San Antonio, Texas, United States*

Dysfunction of biogenic amine neurotransmission is implicated in numerous psychiatric disorders, including depression. The high-affinity transporters for these amines [i.e., serotonin transporter (SERT), norepinephrine transporter (NET), and dopamine transporter (DAT)] are the primary targets for drugs used to treat depression, which include selective serotonin reuptake inhibitors (SSRIs), selective norepinephrine reuptake inhibitors (SNRIs), DAT blockers, as well as drugs that block both SERT and NET, or that block all three of these transporters. It is the increase in extracellular levels of one or more of these biogenic amines that is thought to be critical to initiate downstream events that ultimately lead to therapeutic benefit. However, many patients experience sub-optimal therapeutic benefit or fail to respond to treatment with these antidepressants. The reason for this remains unclear, but it underscores an urgent need to discover new targets to develop novel drugs with improved therapeutic efficacy. We and others have recently identified a previously unsuspected role for organic cation transporters (OCT) and the plasma membrane monoamine transporter (PMAT) in biogenic amine uptake in brain. In particular, we found that when SERT is either genetically or pharmacologically compromised, activity of the OCT3 subtype limits the increase in extracellular serotonin following administration of an SSRI. This finding provides a mechanistic basis, at least in part, for the sub-optimal therapeutic effects of SSRIs: conceivably, uptake by OCT3 prevents extracellular serotonin rising to levels sufficient to trigger downstream events needed for therapeutic effect. Our new data show that co-administration of the OCT and PMAT blocker, decynium-22 (D-22), with either an SSRI or SNRI markedly enhances the ability of these antidepressants to inhibit serotonin and norepinephrine uptake and to produce antidepressant-like activity. Our findings point to D-22-sensitive transporters, putatively OCT3, as novel targets to develop new antidepressant drugs with improved therapeutic potential.

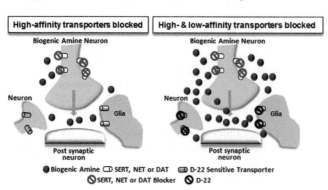

L-DOPA and the Brain Pattern of Dopamine Extracellular Levels: a New Deal Imposed by Serotonergic and Noradrenergic Fibres

Philippe De Deurwaerdère, Sylvia Navailles and Léa Milan

Institute of Neurodegenerative diseases, CNRS UMR 5293, University of Bordeaux, France

BACKGROUND: The therapeutic benefit of L-DOPA is commonly attributed to restoration of dopamine (DA) extracellular levels in the striatum of Parkinsonian patients. The DA effects of L-DOPA overwhelm the striatum favouring the idea that the other monoaminergic neurons, serotonin (5-HT) and noradrenalin (NA), that have a wider distribution in the brain, participate to the release of DA induced by exogenous L-DOPA.

OBJECTIVES: We sought to determine in vivo the involvement of NA and 5-HT fibres in the pattern of L-DOPA-stimulated DA release in the brain of 6-hydroxydopamine-lesioned rats, a rat model Parkinson's disease.

METHODS AND RESULTS: We used multi-site intracerebral microdialysis coupled to high performance liquid chromatography to simultaneously monitor DA extracellular levels in the prefrontal cortex (PFC), the hippocampus (HP), the striatum (STR) and the substantia nigra (SN) ipsilateral to the lesion. Rats received an acute ip injection of L-DOPA preceded by the peripheral decarboxylase inhibitor benserazide (15 mg/kg). Acute L-DOPA (3–100 mg/kg) induced a diffuse and dose-dependent increase in DA release in all brain regions. The increase in DA release was regionally similar at the lowest dose of L-DOPA and 2–3 times stronger in the striatum at the highest dose. A lesion of 5-HT neurons using the 5-HT neurotoxin 5,7-dihydroxytryptamine (intra-raphé), which lowered tissue 5-HT content by more than 90%, prevented L-DOPA-induced DA release in all brain regions.

The administration of the NA neurotoxin DSP-4 (50 mg/kg), which lowered NA tissue content by 85%, enhanced L-DOPA-stimulated DA release in the HP and the SNr. Blockade of NA transporters (NET) using desipramine (10 mg/kg) or reboxetine (3 mg/kg) enhanced L-DOPA-stimulated DA extracellular levels (2–3 times) except the striatum.

INTERPRETATION: These data show that 5-HT neurons are fully responsible for the ectopic release of DA induced by L-DOPA in the brain. Clearance of DA occurs via the NET in areas enriched in NA fibres.

CONCLUSIONS: The mechanism of action of L-DOPA on DA extracellular levels involves 5-HT and NA fibres.

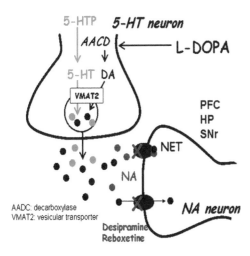

Individual Differences in Nicotinic Acetylcholine Receptor Modulation of Dopamine Signals in the Nucleus Accumbens Shell

Mark J. Ferris[1], Kimberly N. Huggins[1] and Sara R. Jones[1]

[1]*Wake Forest School of Medicine, Winston-Salem, North Carolina, United States*

BACKGROUND: Activation of nicotinic acetylcholine receptors (nAChR) is critical for the acquisition of drug reinforcement and nAChRs are well positioned to govern reward-related dopamine (DA) signals in the striatum. Given the strong relationship between the response to a novel environment and acquisition of drug self-administration, as well as the role of nAChRs in drug reinforcement, we predicted that response to a novel environment would correlate with the magnitude of DA release in the nucleus accumbens (NAc) shell following activation of nAChRs in the ventral tegmental area (VTA) and NAc.

OBJECTIVES: We investigated whether rats with high locomotor responses to a novel environment (HR) would demonstrate greater VTA and NAc nAChR-mediated dopamine release in the NAc shell relative to low responders (LR).

METHODS AND RESULTS: Using *in vivo* voltammetry in anesthetized rats, we demonstrated that HR rats exhibited significantly greater NAc shell DA release following endogenous ACh release in the VTA. Figure 1 demonstrates that DA release following pedunculopontine nucleus (PPTg) stimulation is substantially greater in magnitude and longer in time in HR (A) relative to LR (B) rats, and is sensitive to nAChR antagonists. The differences are characterized by a greater number of spontaneous, transient DA release events (C; $p < 0.05$). Next, voltammetry in brain slices demonstrated that a desensitizing dose of nicotine and multiple nAChR antagonists amplified phasic DA signals in the NAc shell in HR rats, but not in LR rats (data not shown).

CONCLUSIONS: These data demonstrate an enhanced ability of VTA and NAc nAChRs to facilitate DA release in HR animals following both nicotine and endogenous ACh release from PPTg afferents. HR animals are known to acquire drug self-administration more rapidly, suggesting that these animals may have differential DA responses during SA acquisition and conditioned learning that is contingent upon differential modulation of DA signals by nAChRs.

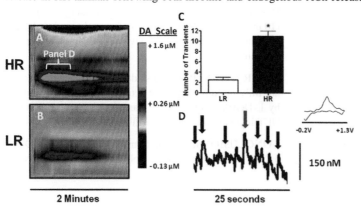

Insulin Induces Long-term Depression in VTA DA Neurons via an Endocannabinoid-mediated Mechanism

Shuai Liu, Gwenaël Labouebe and Stephanie L. Borgland

University of British Columbia, Vancouver, BC, Canada

BACKGROUND: Over-consumption of palatable food can perturb brain reward systems and contribute to feeding disorders. Dopamine neurons of the ventral tegmental area (VTA) represent a critical site for reward seeking and insulin may act in the VTA to suppress feeding.

OBJECTIVES: To elucidate the potential contributions of insulin signaling to neuroplasticity in the VTA, we explored the effects of insulin on excitatory synaptic transmission of VTA dopamine neurons.

METHODS AND RESULTS: Excitatory postsynaptic currents (EPSCs) were recorded from VTA dopamine neurons in mice horizontal midbrain slices. Our previous study demonstrated that exogenous insulin can cause a long-term depression (LTD) of excitatory synapses onto VTA dopamine neurons via endocannabinoid-mediated presynaptic inhibition of glutamate release. In the present study, we confirm exogenous insulin suppresses AMPAR- or NMDAR- mediated EPSCs. This effect did not require postsynaptic intracellular Ca^{2+}, nor was it reversed by later application of the insulin receptor antagonists, S961 or HNMPA[AM]3, indicating that persistent insulin receptor signalling is not required for LTD maintenance. Moreover, the CB1 receptor antagonist, AM251 or the diacylglycerol lipase inhibitor, orlistat did not block the depression when applied after insulin-LTD induction. These results suggest that once insulin-induced LTD has occurred, presynaptic mechanisms maintain the depression. The effect of endogenous insulin under physiological conditions was also examined. Mice were fed sweetened high fat food (SHF) or regular food (RF) for 1 hour. In contrast to RF-fed mice, exogenous insulin did not alter AMPAR EPSCs in SHF-fed mice. To test if this effect was an occlusion by endogenous insulin, we found that AM251 significantly elevated EPSCs in SHF group, but not RF group. Further, mEPSC frequency was significantly less in the SHF group compared to the RF group. Taken together, exogenous insulin as well as endogenous insulin suppresses EPSCs in VTA and this effect is maintained by presynaptic mechanisms.

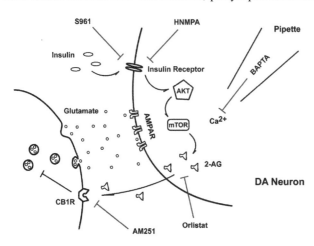